THE FUTURE
OF THE SUN

THE MCGRAW-HILL HORIZONS OF SCIENCE SERIES

The Gene Civilization, François Gros.

Life in the Universe, Jean Heidmann.

Our Expanding Universe, Evry Schatzman.

Our Changing Climate, Robert Kandel.

Earthquake Prediction, Haroun Tazieff.

How the Brain Evolved, Alain Prochiantz.

The Power of Mathematics, Moshé Flato.

JEAN-CLAUDE

PECKER
THE FUTURE
OF THE SUN

McGraw-Hill, Inc.
New York St. Louis San Francisco Auckland Bogotá
Caracas Hamburg Lisbon London Madrid
Mexico Milan Montreal New Delhi Paris
San Juan São Paulo Singapore
Sydney Tokyo Toronto

English Language Edition

Translated by Maurice Robine
in collaboration with
The Language Service, Inc.
Poughkeepsie, New York

Typography by AB Typesetting
Poughkeepsie, New York

Library of Congress Cataloging-in-Publication Data

Pecker, Jean-Claude.
  [*Avenir du soleil*. English]
  The future of the sun/Jean-Claude Pecker.
    p. cm. — (The McGraw-Hill *HORIZONS OF SCIENCE* series)
  Translation of: *L'Avenir du soleil*.
  Includes bibliographical references.
  ISBN 0-07-049182-8
  1. Sun—Congresses. 2. Solar system—Congresses.
  I. Title. II. Series.
  QB520.P4313 1992
  523.7—dc20   91-33142

The original French language edition of this book
was published as *L'Avenir du soleil*, copyright © 1990,
Hachette, Paris, France.
Questions de science series
Series editor, Dominique Lecourt

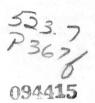

TABLE OF CONTENTS

Introduction by Dominique Lecourt 7

I.   The evolution of the Sun star 21
    The Sun, a Star 21
    Birth of the Sun 26
    Today's adult Sun 32
    Solar and stellar machines 41
    The end of the Sun 49

II. The solar system and its history 57
    The cycle of evolution in the Galaxy 57
    Formation of the planets 60
    New means of observation of the
       Sun from the ground 66
    Role of space research 76
    For continuous observation of the Sun 80
    Cycle and migrations of solar activity 83

III. Astronomy, research, education, pleasure 87
    Research driven by curiosity 87
    Why the Sun and not, for example,
       the Big Bang? 93
    Teaching astronomy 97
    The joy of being an astronomer 101

    Bibliography 105

INTRODUCTION

One of the most intriguing horizons of science is "The Future of the Sun." Should this surprise us? Yes, if science were reduced to a series of inanimate calculations, which it is today popular to disparage; no, if we are really willing to admit that science is a creation of thought, that thought fires the imagination and unceasingly pushes the language to its limits. When the Sun is involved, a world of millennial desires and fears is awakened.

The *Encyclopédie* of Diderot and d'Alembert described that celestial body as "the first object of idolatry" and, ignoring the ancient Chinese, Mayans, and Aztecs, evoked helter-skelter the Baal of the Chaldeans, the Moloch of the Canaanites, the Belphegor of the Moabites, and the Osiris of the Egyptians, ending up with the son of Hyperion, whose praises were sung by Homer, and the god Phoebus more or less identified with the Apollo of Roman mythology. The Sun: symbol of power and omniscience, dispenser of life, agent of death too, a being of flesh and blood. The symbol presented to the Roman legions in 218 as the son of Caracalla, that young fourteen-year-old emperor who assumed the name of Marcus Aurelius Antoninus, later to be known as "Heliogabalus." In 1934 Antonin

Artaud was to find fiery words to celebrate in prose that union of mystical madness of unrestrained eroticism and the fiercest desire for power; a sanguinary madness expressed in defiance of the Universe in its cult of the Sun. The aberration of an unstable mind in a time of decadence? Undoubtedly, but the wise Constantine, creator of the Christian empire in 312, was still to strike medals on which the Sun is represented as his guide and protector. Think, finally, of Daniel Paul Schreber, former President of the Court of Appeals of Saxony, whose case history Sigmund Freud immortalized: he finds in the Sun "the eye of God," transfixing him with its rays, the source of his paranoiac delirium.

The Sun thus holds us with untold subconscious bonds. And the poets have constantly been drawn toward this being associated with such violent symbols, though with infinite precautions. It takes the boldness of Victor Hugo to dream of its extinction:

"The Sun was there dying in the abyss.
The star hidden by the mist, without air to revive it,
Was growing cold, bleak and slowly fading away."

On following the chronicle, marvelously traced by Jean-Claude Pecker, of today's portended death of the Sun, it will be seen that it hardly conforms to the prophecies of the "Death of Satan." Here, the cold is not a sign of death! But the knowledge acquired does

not extinguish the flame of the poets, quite the contrary. Witness the fanciful verse of André Verdet:

"Every star captures in reverie
A Sun in its essence."

It can thus be imagined that the path which led
scholars to recognize in the Sun one star among many,
subject to the physical laws of their common destiny,
was strewn with obstacles.

The ancient Greeks, it is known, devised in a few
centuries a geocentric representation of the solar system, and then of the cosmos, which excluded any idea
of evolution and even of change. There is good reason
to believe that Anaximander (610-547 B.C.), a disciple
of Thales of Miletus (ca. 636-ca. 546 B.C.), proposed
the idea, passed on by Anaximenes (550-480 B.C.), of
the first mechanical model of astronomical phenomena. They all considered the stars to be bodies fixed on
revolving spheres. The idea of celestial spheres was to
make its way through antiquity and the Middle Ages
before being finally shattered only with the beginning
of modern science in the 17th century. The Pythagore-
ans had adopted it, imagining a central fire, hidden in
the bowels of the Earth and surrounded by ten concentric spheres corresponding to the different celestial
bodies observed. Plato (ca. 427-348 B.C.) reaffirmed,
in *Timaeus* as well as in *Epinomis*, the idea of a cosmos

of concentric spheres around the Earth, the movement of the celestial bodies being circular and uniform for each; Eudoxus of Cnidus (ca. 408-ca. 355 B.C.) placed his mathematical genius with incredible success at the service of this bold theory: the movement of the "fixed stars" required only one sphere, those of the Sun and the Moon three and those of the planets four—that is, a total of twenty-seven spheres. And that is how "the phenomena were enshrined"! Aristotle (384-322 B.C.) was to raise this number to fifty-six! Claudius Ptolemy of Alexandria (2nd century A.D.) presented in his *Great Mathematical Syntax*, known in the West by the Arabic name of *Almagest*, a recapitulation and system-atization of all the ancient astronomical knowledge. The same structure, extraordinarily refined, triumphed there as well; that triumph marked the undoing, already accepted for a long time, of the theory of Aristarchus of Samos (310-230 B.C.), the "Copernicus of antiquity," who in the 3rd century before our era had surmised that the Earth moved along the circumference of a circle having the Sun in its center.

What was the nature of the heavenly bodies and, in particular, of the Sun in such systems? Xenophanes of Colophon (570-450 B.C.) had considered the Sun a combination of fiery particles which came together every morning to form a radiant cloud, newborn each dawn. But in the 4th century B.C. the idea that the celestial bodies are really permanent material sub-stances came to the fore. Anaxagoras now affirms that the Sun is an enormous burning stone, since he imagines it to be "larger than all of Peloponnesus." Plato and most of the ancient astronomers consider the heavenly bodies to be igneous in nature. As for Aristotle, he refuses for metaphysical reasons to regard fire as the substance of the spheres. Of divine nature, their circular motion is perfect, that is, eternal, and their incorruptible element is not to be found in "our" (sublunary) world dedicated to "generation and corruption"; what is involved is the ether, a name he gave to that tenuous substance unknown on Earth.

As for the distance of the Sun, the authoritative work on the subject was that of Hipparchus of Nicea (190-120 B.C.), discoverer of the precession of the equinoxes. While Aristarchus had estimated it at one hundred eighty terrestrial diameters, Hipparchus raised it to one thousand two hundred forty-five. That is still nearly ten times less than the distance currently accepted.

With his celebrated *De revolutionibus orbium coelestium* [On the revolutions of heavenly bodies], Nicolaus Copernicus in 1543, the very year of his death, placed the Sun at the center of the Universe. "At the center of everything," he writes, "resides the Sun. Who in that splendid temple would put that lamp in a better place, whence it can illuminate everything all at once? It is not wrongly then that some call it light or soul or master of the world, that Trismegistus names it the Visible God or the Electra of Sophocles, the All-seeing. For, as if seated on a royal throne, the Sun governs the twirling family of the stars." Already, but in a literal astronomical sense: King Sun! And by divine right. It would have to be dethroned for its true nature to be prosaically laid open to scientific investigation and for the question of its history to take on an objective meaning.

Without doubt, the one who was to contribute most to that end was, paradoxically, one of Copernicus' most fervent admirers and among the most mystical as well: the great Johannes Kepler (1571-1630), who saw in the Sun the expression of God the Father; in the stellar system, that of the Son; and in the light and force circulating between the two in space, that of the Spirit. But Kepler formulated the true laws of planetary motion and dislodged them from the geometric center of the system where Copernicus had installed it. Albeit still vaguely, he advanced the first hypothesis of a dynamics associated with the Sun, which one must be careful not to interpret as an "attraction," and though he rejects the idea of an infinite Universe, he pushes back the limits of the Copernican system, already so much vaster than that of the ancients.

It was, however, with the observations of Galileo (1564-1642) that the decisive threshold was crossed. His *Siderius Nuncius* (The sidereal messenger, 1610) reports the extraordinary discoveries made since 1609 thanks to the use of a telescope manufactured by the Murano glassmakers: lunar mountains, satellites of Jupiter, phases of Venus, resolution of the cluster of Cancer into stars and.... sunspots.

The telescopes were enthusiastically aimed at the skies. And very soon the sunspot phenomenon attracted attention. It was certainly not a question, properly speaking, of a discovery, since the first mention of such spots dates back to Theophrastus, who lived in Athens between 372 and 287 B.C. The annals of China and of Japan also report observations of the same kind since 28 B.C. As important as was the work of the Dutchman David Fabricius in 1611, Galileo was the first to undertake a systematic study of sunspots, followed shortly thereafter by Father Christoph Scheiner, a German astronomer (1575-1650) of Ingolstadt with whom he had a lengthy controversy over interpretation of the facts.

The idea of a solar "activity" thus makes its appearance, and the question of its matter is posed in new terms: Descartes, in *Principia philosophiae* [Principles of philosophy, 1644] writes: "The Sun has in common with fire and the fixed stars the fact that the light emanating from it is not borrowed from elsewhere," and concludes that it is therefore composed "of a highly liquid matter, the parts of which are so extremely agitated that they carry with them the portions of the sky adjoining and surrounding them."

From the distribution of these spots, their migrations and their regularity, Galileo drew another lesson, which also opened up a whole line of research for him: the Sun is driven in a rotating motion; and, as Scheiner showed, it does not turn like a solid globe: the equatorial regions revolve much faster than do those at high latitudes. Descartes would find an argument there for his theory of the fluid Sun. For all that, the spots had still not yielded all their secrets. As we shall see, they were to point the way to the magnetism of the Sun and of the solar cycle.

The question now arose of accounting for the formation of the solar system through the laws of physics. Descartes, in his *Traité du monde* [Treatise of the world], published fourteen years after the author's death, was the first to pose the question, with the caution duly demanded with regard to biblical theses. Strictly assimilating matter to space, Descartes rejected the existence of a vacuum as strongly as had Aristotle, though for other reasons, and thus attributed the origin of the solar system to the movements and encounters of great vortices of matter derived from an initial chaos. The given facts of the problem were to be turned topsy-turvy by the Newtonian theory of universal gravitation (1687): the cartesian vortices, objects of sarcasm, would become things of the past. Careful observation of comets, those celestial bodies previously considered capricious and unpredictable, which Newton had somehow integrated into the solar system by subjecting them to the law of attraction, was to guide prevailing thought. Buffon (1707-1788), in his *Théorie de la Terre* [Theory of the Earth, 1749], believed that the Sun and the comets antecede the planetary system. He attributed its birth to a catastrophe: the collision of a comet and the Sun! Reviewing the question thirty years later in *Époques de la nature* [Periods of Nature, 1778], he presented a detailed explanation of this origin of the solar system and asserted that all of the matter of Earth and the planets was torn away from the Sun.

The *Exposition du système du monde* [Exposition of the system of the world], a work published in 1796 by Pierre Simon Laplace (1749-1827) before the appearance of his great 5-volume synthesis, *Mécanique céleste* [Celestial mechanics, between 1799 and 1825], propounded quite another hypothesis: that of the formation of the solar system from a progressively cooling primitive nebula. Unknowingly, Laplace was echoing a thesis advanced on a purely speculative basis by Kant in his *Theory of the Heavens* (1755). Hence, the name Kant-Laplacian hypothesis is often given to this "nebular" hypothesis.

The thinking of cosmologists has ever since been constantly swinging back and forth between catastrophist views and revised "Laplacian" views. It seems that today the latter have the best arguments to offer. Those arguments can be traced essentially to the progress of observation due to the improvement of instruments and to the establishment of astrophysics as a science.

Photography had already enabled Hippolyte Fizeau (1819-1896) and Léon Foucault (1819-1868) to make the first daguerreotype of the solar surface in 1845; and Jules Janssen (1824-1907), founder of the Meudon Observatory, took daily photographs of the Sun starting in 1875. But new knowledge would come essentially from spectroscopy, which made it possible to "explore" the interior of the Sun and of the stars. Astronomical spectroscopy can undoubtedly be dated back to the discovery by the English physicist and chemist William Hyde Wollaston (1766-1828) of the dark lines which criss-cross the solar spectrum, an observation explained and interpreted by Joseph von Fraunhofer (1787-1826) in Munich in 1811. But spectrum analysis was really founded by Gustav Robert Kirchhoff (1824-1887) in 1859, with the aid of celebrated chemist Robert Wilhelm Bunsen (1811-1899). Kirchhoff demonstrates that the "dark Fraunhofer lines are formed in association with the atoms of the solar atmosphere; during an eclipse those atoms alone emit observable radiation, and their spectral lines appear brighter, without the incandescent radiation of the surface hidden from our eyes by the lunar disk." In 1897 Rowland was to make the first photograph of the solar spectrum.

16

INTRODUCTION

Sir William Huggins (1824-1910) and Sir Norman Lockyer (1836-1920) compared some stellar spectra with laboratory spectra in great detail. The Italian astronomer Angelo Secchi (1818-1878) and then Edward C. Pickering (1846-1919), at the Harvard College Observatory, undertook to classify the spectra of stars. Secchi ranked stars in four categories (white, yellow, red, ruby). Our Sun is in the second category. The layers of the solar atmosphere were explored, replete with strange and active phenomena. The American astronomer George Hale and the French astrophysicist Henri Deslandres (1905) made the first "spectroheliographs" available to the Mount Wilson and Meudon observatories, respectively. Later, Bernard Lyot (1910) invented the "coronagraph," which affords the opportunity for permanent observation from the Pic-du-Midi Observatory of the solar corona, previously visible only for a few minutes every two or three years in the privileged places where a solar eclipse can be seen.

It is also thanks to spectroscopy that the first theoretical analyses of the Sun's atmosphere were made by A. Schuster (1851-1934) and K. Schwarzschild (1873-1916) in 1905 and 1906. Though knowledge of the Sun and the stars had thus made tremendous progress thanks to a new technique, it was again going to be revolutionized in the early decades of this century by the progress in physics. The turning point came with the discovery of nuclear reactions controlling the energy flow of stars. British astronomer Sir Arthur Stanley Eddington (1882-1944) played a pioneering role here. His work entitled *The Internal Constitution of the Stars* (1926) presented a detailed model of the stars' interior and described the behavior of the atomic particles constituting it. Carl von Weizsäcker in 1938 and then Hans Albrecht Bethe in 1940 contributed the first precise information necessary for an understanding of the carbon cycle and hydrogen reactions.

This work opened the way to a new conception of stellar evolution—a conception that was to be steadily broadened up to our time, expanded and reinforced by observations from space and the use of computers.

This is the fascinating story that the reader is going to discover, written by one of the undisputed masters on the subject, heir to the great French solar tradition which flourished from 1870 to 1935 and was distinguished by so many great names working at those "magical places" like the Meudon and Pic-du-Midi observatories, in particular. It is that tradition which inspires the host of young researchers who are making it possible for France to play a major role internationally; to be on the cutting edge of solar system research. In reading the pages that follow, the reader will discover the impressive projects they have started or to which they have contributed.

Jean-Claude Pecker pleads with passion for a science "driven by curiosity." He highly acclaims the sheer joy he experiences on observing the sky and acquiring knowledge of the heavens. Who will say, after having read his words, that science makes the world "a less enchanting place"?

Dominique LECOURT

I

THE EVOLUTION
OF THE SUN STAR

THE SUN: A STAR

The idea that the Sun is just one star among many, similar in nature to the others, was very hard to accept. This, in fact, is not at all evident on observation: what is immediately striking, rather, is the uniqueness of this celestial body, the visible differences in size and luminosity which distinguish it from all the rest; the fact that when it is "glowing" the brilliance of the others fades. Hence, those so very beautiful and numerous myths which deified the Sun at the dawn of human civilization, among the Incas, the Aztecs and the Mayans, in Egypt, in Greece and in the Orient. The intuitive perception that the Sun and the stars are celestial bodies of the same nature was well established among some of the ancients and in the works of Johannes Kepler (1571-1630), but still it was only an intuition, as it was to be for Giordano Bruno (1548-1600) or Galileo (1564-1642). For proof to be furnished, it was first necessary to be able to shake loose misleading appearances and to show that the absolute luminosity of this body is of the same order of magnitude as the absolute luminosity of the stars. In other words, it was first necessary to be able to determine the distance of a given star and say: "If I placed the Sun at the distance of that star, their brightnesses would be roughly comparable." It would have to be possible to correlate the radius, mass, luminosity and color of the Sun and the stars.

Now, such a demonstration and such measurements were only provided, in two successive stages, as late as the 18th and 19th centuries!* We thus know today that the Sun is close to us: not more than 150,000,000 km [93,000,000 mi] distant, while all the other stars are light-years away: this is *our* Sun. It bathes us with light and heat. All of the energy we possess on Earth originates more or less directly from the Sun. It has been suspected for a long time that its activity is not without influence on the climate prevailing on Earth. We cannot then ignore a celestial body on which our very existence depends. But another interest, of a theoretical nature, follows from

---

* The distance of the Sun was precisely determined by studying, from the entire surface of the Earth, the passage of Venus in front of the Sun in 1761 and 1769. Then, the distance of the nearest stars was deduced from the apparent dimension of their displacement in the course of the year, against the background of very distant stars (a dimension on the order of some tenths of a second of arc!): T. Henderson (1798-1844), F. Bessel (1784-1846) and F.W. Struve (1793-1864) made these measurements independently of each other at the Cape of Good Hope, in Germany and in Russia, respectively, in the 1830s.

what is, after all, a commonplace star: by studying it, we can draw conclusions about the other stars. Indeed we cannot overlook the fact that the mass, radius and luminosity of barely fifty stars of different types have been determined to date with precision and by direct methods. Learning about the burning heart of the Sun, the layers of its atmosphere and the wind that perpetually snatches matter from it means opening a window of knowledge of the other stars.

Once it was accepted that the Sun is a star (a star whose proper name is Sun), it became necessary to resolve the question of the origin of its radiation—a question that has been asked all the more emphatically as the chronology of Genesis was more definitely abandoned. If, in fact, instead of ascribing to the Sun (and to Earth) an age of 4,000 to 6,000 years, we reckon it in billions of years, as we do today, it is difficult to understand how its radiation has not already been exhausted. The hypothesis had originally been formulated that a chemical-type reaction, say a carbon combustion, was involved. The Sun would have been a sort of great incandescent coal! But on such time scales, that hypothesis was no longer tenable. Lord Kelvin (1824-1907), who had in mind for geological phenomena a scale that was very long, but relatively short by comparison with what we accept, went one step further and tried to explain the duration and quantity of radiation by the slow contraction of the Sun. But when it was discovered that it was necessary to count in billions of years, another type of explanation had to be considered. The great French physicist Jean Perrin (1870-1942) was the first to state, in the 1920s, that it is probably the transformation of four hydrogen nuclei into one helium nucleus with loss of mass which supplies the energy. A portion of mass disappears and a portion of energy appears in the form of radiation according to the well-known Einstein equation: $E = mc^2$.

That general view still remained overly qualitative, however. If thermonuclear reactions could actually take place, it was necessary to explain them in detail and to show their precise effects in order to account, notably, for the temperature and density of the Sun.... That demanded far more progress in determining the detailed physics and mechanics of nuclear reactions. What could not be accomplished by Jean Perrin, given the state of science in 1920, was later achieved by physicists like Hans Bethe. It is worth noting in this connection the process of interaction which then ensued between astrophysics and laboratory physics (but had it not existed since Bacon or Galileo?). The work of Bethe and von Weizsäcker on solar energy was a powerful stimulus to the study of atomic nuclei and of the probability of a given reaction. Particle accelerators and theories of nuclear structure owe a great deal to the study of the Sun!

But now that the production of energy was associated with the transformation of hydrogen atoms into helium atoms, the question of the evolution of the Sun again arose. Since it could not be regarded as an infinite mass of matter, it had to be assumed, in particular, that this nuclear fuel would one day disappear. There was a time when it had "to have been ignited": how did that come about? There would come a time when it "will be extinguished": how and when would that happen?

It was possible to answer these questions only after having carefully studied the thermonuclear reactions taking place in the very hot interior regions of the Sun.

But, in general, when a star is studied, it is necessary to know the conditions under which it appeared in the midst of a diffuse interstellar medium, and to determine the time when it was formed as such and began to radiate; one must then study the "adult" star, during the period when it radiates regularly and, finally, in the period when the hydrogen begins to be depleted in the hot regions and when instability phenomena are activated, ending in catastrophe....

Such a study must of course clearly differentiate between phenomena having a large-scale evolutionary significance and those that do not. This often proves difficult: in the case of the Sun, for example, it had for a long time been postulated that sunspots had such significance, revealing some phenomenon of secular evolution. Sunspots were, in fact, the first known manifestation of its activity. They have been observed, counted and measured ever since Galileo and the German astronomer Scheiner. It was realized that these spots (the size of each is reckoned in tens of thousands of miles, many of them being larger than Earth) do not always appear. During some periods, between 1630 and 1690, for example, none was observed. A certain periodicity, it will be seen, marks their manifestations. It was, finally, perceived that they are not equally distributed over the solar surface, appearing at well-defined latitudes, but rarely higher than 45°. Their long duration (one spot lasts several months!) was used in the 17th century as a way of proving and measuring the rotation of the Sun. But as for the significance of these spots for the evolution of the Sun, a great deal of time has been squandered searching for it; when we will have understood the overall processes of that evolution, then will be the time to relate particular phenomena to it, such as sunspots and the active regions surrounding them, and to analyze their characteristics as complex symptoms of the physics governing the solar machine.

BIRTH OF THE SUN

Let us then turn to the initial medium in which the Sun was formed as a star. That medium is the Galaxy—our galaxy, one among billions: a disk of approximately two hundred billion stars, slightly bulging at its center, where the density of stars is very great. The Sun occupies an almost peripheral position on that disk; when we look in the plane of that disk, we see the Milky Way, which is a sort of outline of the Galaxy in the sky. The Milky Way crosses the entire sky, from one end to the other. It is the "great circle" of the sky; but from our almost outside position, we have a good view of only the densest central regions, those of the constellation Sagittarius. The Milky Way is the Galaxy, as seen by us on the edge, one might say. Thanks to what are known as "fisheye" lenses, which photograph the whole sky, it is possible to see clearly the bulge of the Galaxy at its center; and it then also becomes very clear that this Galaxy resembles other galaxies outside our own. It is known that millions and millions of such galaxies exist, near and far. Our Galaxy is made up, we are told, of two hundred billion stars. Between those stars one finds interstellar matter, most of it cold. When that matter is cold, it is composed of ordinary (nonionized) atoms, but also of very numerous molecules and more or less coarse grains of dust.

In fact, what we have just mentioned is one of the most important discoveries of the last thirty years, for it has been demonstrated that, among those molecules, some are also found in living matter; they are already complex molecules, objects of the most classic form of organic chemistry (formic acid, ethyl alcohol, etc., but also benzene molecules).

The interstellar molecules are observed in a free gaseous state; but they are also present on the surface of the grains of dust where they form. The matter of the cold regions (approximately −150° to −200° C [−238 to −328° F]) of interstellar space thus consists of grains of dust, composed essentially of graphite, silicates, and cold gases. It is amidst that dust, within those gases, that a star may later form. It should be added that there are also fairly hot gaseous zones (approximately 10,000°C or 18,000 °F) in the vicinity of hot stars; owing to the existence of a hot star (surface at 20,000-30,000°C or 36,000-54,000°F), a whole medium is, in fact, created, ionized and heated by the high-energy radiation of that star (ultraviolet radiation). The grains of dust are then partly destroyed, as are the molecules, and the atoms are ionized. It should also be mentioned that actual bubbles, even hotter and almost empty, exist here and there in the Galaxy.

In that initial medium, under the effect of shock waves (which can be provoked by the explosion of a supernova, by phenomena localized around certain stars of the Milky Way or even by the passage of a spiral arm of the Galaxy—the Galaxy has a spiral structure), a condensation of matter can take place. That condensation can be considered a sort of primer, a germ, giving rise to a fairly rapid collapse of the matter. What happens then? Lumps are formed within the matter. Such a lump is relatively cold, when its temperature (a few hundred, and then a few thousand degrees) is compared to that prevailing in the center of a star. This "low" temperature prevents any effective nuclear reaction, so that the lump, carried along by its own weight, collapses while broadly speaking retaining its mass. It is then separated from the rest, essentially maintaining the same angular momentum: it is driven by a rotary motion arising out of the initial turbulence of the galactic medium; that rotation then accelerates as the lump condenses. One can say, by way of simplification, that the angular momentum is the product of the angular velocity of rotation and the size of the object: in condensation, that product remains constant. Now, the size of the lump diminishes, since it collapses; the velocity of rotation increases, until it reaches velocities close to those observed in the stars.

During collapse of the lump, a number of phenomena occur, including probably the formation of planets and an increase in temperature. This is understandable enough: as the mass collapses, its potential gravitational energy diminishes, being transformed into kinetic energy and then into heat; that is exactly what happens when energy is liberated by an object falling to the ground. Once there is rotation, the condensation process is not as simple as we have just described it. The nebular matter can accumulate along the equator of the star being born, forming an "accretion disk," while at the poles, on the other hand, stellar

THE FUTURE OF THE SUN

evolution can entail the ejection of rapid streamers of matter. Those phenomena are clearly visible on the young stars in an early stage of formation called "T Tauri stars" (named after the T Tauri star, the first of that category to attract attention). Did the young Sun behave like a T Tauri star? We do not really know. But those complex processes, like the ones associated with the evolution of the magnetic field "trapped" in the matter in condensation, clearly demonstrate to us that we should not be content with vague theories: those evolutions are established by elaborate calculations, and only modern supercomputers are capable of simulating a secular evolution in just a few hours, using the demanding and exact equations of physics.

The energy liberated during condensation escapes in the form of light. The medium is heated, the heated medium radiates, and this radiation causes it to lose energy! In this first phase the temperature then increases, but that increase is almost exactly compensated by energy leakage; the evolution remains moderated, in spite of the swiftness of collapse.

A moment then arrives when the density of the lump has become much greater and the medium is more and more opaque. A kind of struggle then ensues between that growth of opacity and the natural propensity of the energy produced in the center to dissipate. The opacity prevents light from escaping as readily as before, so that the temperature now rises very rapidly. The lump is no longer transparent: it truly becomes a star, and remains stable because its increase in temperature is almost exactly compensated by the escape of light. The birth of a star can thus be traced to the day when the lump is no longer transparent. A star is then a very dense region of the Universe submerged in a region that is almost a vacuum. As I have previously written, "its whole history is one of the boundary between that plenum and that vacuum."

Accordingly, at this point in the narration, the temperature is then still too low for nuclear reactions to be kindled. The star thus born then continues losing energy; but its collapse is now retarded, its contraction having been slowed down.

After a certain time, however, the central regions of the star become very hot, because of how difficult it is for the radiation to escape. Their temperature reaches ten, twelve and even fifteen million degrees. Thermonuclear reactions will then be able to start. Those reactions depend, of course, on the temperature prevailing at that time. Very quickly they compensate the effect of opacity. The captive energy is excited! And, finally, the star reaches a true state of equilibrium: the central temperature then attained establishes the type of nuclear reaction which takes place; those conditions are controlled by the mass of the star. If the star has a large mass, the central temperature will be higher than if it has a small mass. If the central temperature of the star is low, proton-proton type reactions prevail; if the central temperature is higher, other reactions occur involving carbon, oxygen, and nitrogen nuclei (those of "Bethe's cycle"). When all is said and done, in the case of both these types of reactions, four hydrogen atoms are transformed into one helium atom, and the higher the central temperature, the faster this transformation.

But here, even in the case of the Sun, the data of physics are still insufficient for the results to be completely accepted, and the debate continues apace. In the mid-1980s, for example, scientists were convinced that the temperature at the center of the Sun was fifteen million degrees; at present, after a recent revision of data used in calculating the rates of nuclear reactions, astrophysicists lean toward a somewhat lower temperature of about thirteen million degrees.

TODAY'S ADULT SUN

Here we are then, having arrived at the stage of an adult star like the Sun, which no longer draws its energy from the initial contraction it underwent, but from nuclear reactions. The Sun, let me point out, has 4.5 billion years behind it. It has already consumed an enormous quantity of its central hydrogen, but still has a great deal left. In short, the Sun is at its midway point or, dare I say it, at high noon in its life. Astrophysicists construct models of this adult star. By "model" they mean a description of all the physical quantities that characterize it. Yet the very concept of model remains somewhat ambiguous.

There are actually two types of models. The first consists in gathering the greatest possible number of observations made and then attempting a rational extrapolation from them with the aid of the equations and laws of physics. But the observations relate only to the light or particles which, emanating from the Sun, have reached the observatory receivers. Consequently, they are incomplete; thus, we shall never see the visible light as emitted in the center of the Sun, since it is absorbed, emitted, reabsorbed and re-emitted billions of times between the moment it is created by a thermo-nuclear reaction and the time it reaches our eyes. We can truly say that it is then no longer the same light; it has not retained the memory of its birthplace. True enough, it has totally preserved the *quantity* of energy it has received, but the *quality* of that energy has been much "degraded" since the time of its production; it has drifted to the long wavelengths representing low-energy radiation. The quantity of energy that Earth receives, then, really constitutes a total measure of the processes unfolding at the center of the Sun; but that measure remains very indirect. This type of model (empirical) is then aimed primarily at determining the structure of the regions of the Sun, from where the light observed or the cosmic particles of solar origin ema-nate; that "diagnosis" can extend to the interior regions of the Sun only through reasonable extrapolation, based on the equations of physics, for which the empirical models provide boundary conditions.

It is then possible to envisage determining in advance the phenomena which will be observable in the future thanks to, for example, a technical expansion of exploration opportunities. The new observations will serve to test the empirical model; the quality of extrapolation and the accuracy of the physical equations used will be put to the test. And many surprises have often already turned up!

But there is another type of model, the elaboration and finality of which are very different. In its construction, only a minimum number of observations regarded as characteristic is retained. It is said, for example, that the Sun is a star with a mass of two billion billion billion tons, or 2×10^{33} grams, and that it is a star having a given luminosity, a given size, perhaps a given speed of rotation, and a given total magnetic field. But that is as far as it goes! Based on these few facts, a model is constructed that is perfectly coherent from the standpoint of physics, hence introducing as few empirical data as possible. The quantities of mass, radius and luminosity will then serve as boundary conditions for the equations. What would more correctly be called an "archetype" is thus obtained, and it is then tested with the sophisticated physical theories born of the imagination of physicists. This involves quite another type of modeling which, when the opportunity arises, makes it possible to understand many things and has the merit of physical consistency; but observational verification of any findings is not always easy or even possible.

One might say that the scientific mind takes different approaches to these two types of models. It is very uncommon, in fact, for the same individual to use both. There are thus observers who gather empirical data of all kinds, whether in space or on the ground, who dedicate themselves to following the progress of the Sun twenty-four hours out of twenty-four—because it is actually changing all the time. This type of research proves particularly attractive in the case of the Sun, considering its proximity: it is possible to scrutinize the solar surface in its finest details (50 miles!), to make successive observations within a fraction of a second, to enhance precision, to increase the resolving power of the instruments used, and to subject even the most minuscule phenomena to analysis. What cannot be done with other stars, because the quantity of radiation we receive from them is too small, can be done with the Sun, a star within our reach!

But coexisting with these precise observers and with the modelmakers of such phenomena are "modelmakers" of the second type, who are little concerned with observations, whose work appears to be essentially theoretical, based on very complex calculations, which modern data processing invests with extraordinary power and precision.

For theoreticians as well as for observers, these two approaches clearly do not rival but rather complement each other. We are going to illustrate this idea with a detailed example.

A true model of the interior of the Sun cannot be constructed with the first method, since, as we have stated, it is not possible to observe it directly. It is possible, on the other hand, to construct models of the solar atmosphere, that is, of the region from which photons, or light quanta, directly originate to reach the observer.

The Sun is darkened on its edges: the rays of light come towards the observer from the center of the solar disk, crossing the layers perpendicularly; but, on the edge of the Sun, those rays are inclined in relation to the outer layers; they thus originate from shallower layers. The fact that the disk is darker on the edges of the Sun shows that the temperature is lower there than in the central regions emitting the radiation. This roughly means that, since the center represents deeper layers than the edges, the temperature increases as one penetrates deeper into the Sun. But this is far from evident on simple observation, even though it can today be confirmed in many other ways.

Now, on the basis of that simple observation— darkening at the edges of the Sun—it is possible to deduce the distribution of temperature according to depth in the layers responsible for radiation. But many other things can also be learned about the composition and structure of that star! It had been known for a long time that a great deal of hydrogen is found in the Sun's atmosphere. But only through study of the darkening at different wavelengths of the spectrum was it possible to deduce that the principal absorbent in the solar atmosphere is what is called the "hydrogen negative ion," an atom which does not exist on Earth and consists of one proton surrounded by two electrons instead of only one. The existence of that negative ion had been theoretically established by physicist Hans Bethe in 1929 on the basis of work in quantum mechanics. Rupert Wildt, a German-American astronomer, had shown in 1940 that it could be found in the solar atmosphere where he indicated that it might play the role of principal absorbent. But it was not until 1949 that Daniel Chalonge and Vladimir Kourganoff, working in France, were able to show empirically, through analysis of the solar spectrum, the absorption spectrum of the negative ion. That discovery could not have been made on the basis of pure theory; it implied the use of an empirical model.

Conversely, pure theory, being founded on a solar model (in the second sense of the term), has notably advanced research on the convective motions affecting the Sun. Take the most commonly accepted solar model. It shows that convective activity must necessarily develop at a certain depth: a convective zone would extend between a layer situated at approximately 0.7 solar radius (which is 696,000 km or about

435,000 mi) from the center of the body and a layer much closer to the apparent surface, a few hundred miles from that surface. Only through purely theoretical means has it been possible to calculate the location and thickness of that convective zone. A great deal still remains to be done, however, to arrive at a satisfactory result! The indirect observations available to us show, since they are not concordant, that considerable progress remains to be made. To what are these uncertainties to be attributed? Undoubtedly, to the absence of means of empirical observation: this theory had been developed for a very long time without any means of examining the movements affecting the interior of the Sun other than the "rice grains" (or "granulation") on the solar surface. Now, an almost direct means of exploration of deep convection has recently been discovered. The Sun, it has been observed, oscillates at every point; it is agitated by movements that can be described as "seismic": waves cross its mass and, after propagating there, are reflected on the solar surface. Those waves, like the seismic waves that enable us to probe the depths of the Earth, provide us with information about the depths of the Sun. By studying those global oscillations and comparing them to the results that would be obtained from the model of the convective zone, the theoretical model of convection has thus had to be slightly corrected in the last fifteen years. Describing convection is, in fact, not a simple physical problem. If the study of solar oscillations (known as helioseismology) has uncovered contradictions and made it possible to improve the model of the inner Sun, it is to the extent that the theoretical studies of convection had been founded on approximate methods described as "phenomenological," insofar as they are inspired by an intuitive view of phenomena, rather than by a detailed and painstaking examination of the behavior of physical equations, a meticulous examination still rendered impossible by the difficulty of those equations.

To fully understand the importance of this research, one must not lose sight of the intuitive phenomenological idea of what is called the convective zone. In a star—and, therefore, in the Sun—what is involved is a region in which, if a movement of matter starts upward, it must necessarily continue to the top of the zone considered; if it starts downward, it will continue until it reaches the bottom of the zone; a general mixing thus takes place within the convective zone.

A mental picture (insight is what made possible the theory we have discussed) helps explain the nature of these convective motions. Let us isolate in thought a small bubble of gas, imagining that it rises sightly and remains separated from the surrounding medium. Its temperature and density will then be modified in an "adiabatic" manner well known to physicists. But, taking that modification into account, the bubble becomes lighter than the surrounding medium if the ambient density does not decrease too fast with altitude. The bubble will thus continue to rise; if it had become denser that its environment, it would have redescended. Hence, depending on whether the density gradient in the star is greater or less than the adiabatic gradient, convection will or will not take place. If density decreases faster than the adiabatic movement implies, convection stops. A "radiative" zone is then entered, where the convective movements can no longer occur. In the Sun, as we have stated, the central regions (outside the very hot ones where thermonuclear reactions are produced) are essentially radiative up to 0.7 solar radius; a convective zone begins there, extending almost to the surface; a thin new radiative zone is then encountered. Outside, everything becomes very complicated, as will be seen presently, with the solar wind and corona.

Is what we observe with respect to the Sun—namely, the existence of distinct "layers" and of a convective zone—true of all stars? Yes and no. By seeking to universalize the idea of a "convective zone," do we not run the risk of being victims of extreme heliocentrism, due to the fact that the Sun, being closest to us, is the star with which we are most familiar? The best way to guard against such error is to consider the extreme diversity of the stellar medium, as revealed to us by the now highly exact studies conducted on a very large number of stars. It is then evident that the convective zone greatly depends on the luminosity of the star: significant in some cases, nonexistent in others. Its depth also depends on the type of star studied. These theories, and these results, are thus tentative; as, in the case of the Sun, very imperfect models are involved, founded on a phenomenological theory, and the seismology test still only very rarely applies. And, although theoretical models are devised for each type of star, there are still only about a dozen in which oscillations of a type comparable to those observed on the Sun have been detected. Such are the "solar models": empirical and theoretical and, as in the latter instance, corrected by observation. It is now apropos, before approaching the question of the Sun's evolution, to provide some details on the solar machine itself.

SOLAR AND STELLAR MACHINES

For a long time, until the 17th century, it had been believed that the Sun was a solid or liquid body. It is now known that it is a ball of gas. Empirical evidence is available of the increase in temperature toward the interior of the body, and theoretical justification of that increase has also been adduced. In principle, matter can exist in several forms: liquid, solid, gaseous. In practice, its state is defined by temperature and pressure. Above a given temperature and a given pressure in a state of equilibrium (called "critical point"), matter can exist only in gaseous form. Now, the temperature of the visible solar surface is about 6,000°C [11,000°F]. Consequently, the hotter interior can consist only of a gas. All of the theories and models of the Sun are thus founded very consistently on the idea that the Sun is a perfect ball of gas; in a "perfect" gas, like monatomic laboratory gases, the pressure is proportional to the temperature and density. Stellar modeling is based on the same principle.

The pioneering work on this subject is that of Robert Emden (1907), *Gaskugeln* (Gaseous spheres), a theoretical discussion of the equilibrium of sphere-shaped gaseous masses. This work opened the way for numerous other studies which dominated astrophysics until some twenty or thirty years ago. But, apart from the fact that the energy source could not be identified, this theory and these studies ignored three important facts.

To begin with, as has been seen, the Sun is a rotating body, like all stars. Next, convective activity plays an important role in channeling, not only energy, but "momentum," that is, the quantity of motion. Finally, the Sun, being a magnetic body, is a giant magnet.

Those three features make the Sun a machine, and that machine is a dynamo. Actually, the observational aspects of that machine had been noted for over two centuries. Think, for example, of Galileo observing the sunspots: since the 18th century it has been known that these spots were associated with a significant magnetic field. And since that time it has also been known that those same spots could be used to measure the solar rotation-producing currents, which in turn generate magnetism. It is recognized that this rotation is slow, taking somewhat less than one month. It is not uniform: more rapid at the equator (one revolution in 25 days) and slower at the pole (one revolution in 30 days). That phenomenon is known as "differential rotation." Finally, it was clearly realized that solar granulation is a surface manifestation of deep convective activity.

In the 1850s Jules Janssen (1824-1907) discovered the Sun's "rice grains," or "granulation." Perfected and clarified in the 1950s by the German-born (1912) American astrophysicist Martin Schwarzschild and by Jean Rösch and Constantin Macris at the Pic-du-Midi Observatory in France, those observations revealed that the granules, somewhat hotter than the neighboring gaseous medium, ascend and divide into smaller granules. Schwarzschild had suggested comparing the physics of this phenomenon with that of the cells formed in a pot of water about to boil. Today, it has been established that all these phenomena correspond to coupled processes: a permanent play has settled in among magnetism, rotation, and convection. Differential rotation twists the lines of force of the magnetic field until unstable states of tension are reached. The solar activity cycle, although known for two centuries, can now be better understood; during periods of maximum activity the magnetic field is strong and resembles a multipolar field. During periods of minimum activity, it resembles rather a bipolar field. This initially simple bipolar field twists, retwists and complicates differential rotation to the point of a maximum growth of activity linked to that magnetic complexity. Then, with instability reaching an excessive level, everything is simplified little by little, the magnetic energy is released thanks to the eruptions; the Sun gradually recovers its quiescence. It thus seems that there is a kind of alternation between the "poloidal" structures (bipolar, like those of a magnet) and the toroidal magnetic forms (the field then affecting the form of rings winding around the Sun).

Given all that, our theoretical knowledge of this extraordinary machine is still in its infancy—an infancy that looks as though it might be prolonged, because we have almost no information on the other stars that would enable us to vary the parameters as they affect the Sun. These parameters are mass, luminosity, radius, chemical composition, rotation, and (average) magnetic field. If they are varied, we are talking about a different star. For lack of actual experimentation, as is always the case in astronomy, we would have a sort of quasi-experimentation. The effect of variations in each of these parameters on the others could thus be determined. For example, to determine the effect of luminosity of the star on the active phenomena, one can study a somewhat hotter star and analyze the active phenomena occurring there. But, as has been seen, we are still very far from being able to conduct such "experimentation" successfully; little is known about stellar activity in spite of promising observations. However, we have been able to obtain sufficient data by this means to accumulate a body of doctrine on the general evolution of stars; this doctrine is based on a small number of parameters; it takes very little account of magnetic activity phenomena, but does prove coherent and productive.

A famous simple theorem exists, which has been widely used in astronomy. Dating back to 1926-1927, it is known as the Vogt-Russell theorem, after Heinrich Vogt and Henry Norris Russell, and it states the following: a star is totally defined when its mass and chemical composition are known. This theorem, as is often the case with famous theorems, is valid only on first approximation. It does not take into account, for example, the fact that two stars can have the same mass, the same chemical composition and yet different speeds of rotation. Though false, it has proven very useful, however, because the schematization it proposes is immediately applicable. It is known that stars grouped in clusters exist in the Galaxy; when stars are formed, they can, in fact, be arrayed in families. When these families contain only a small number of stars, they tend to be dispersed, in which case we speak of an "open" cluster. An open cluster is generally young, for—if it were old—it would have had time to disperse. If, on the other hand, it is composed of a large number of stars, they often remain crowded together, held close to one another by gravitational forces which, as it were, play the role of cohesive forces of the cluster. Such a cluster therefore has a spheroidal shape and is said to be a "globular cluster." The Pleiades, which are visible to the naked eye and were catalogued by Charles Messier (1730-1817), are a typical example of an open cluster, and the Hercules cluster is a typical example of a globular cluster. Star clusters (whether open or globular), then, are composed of stars born together, having the same chemical composition but different mass. Now if I try to describe their properties, I must find a kind of "one-to-one" relation between the mass and each of the measurable properties of these stars.

That is precisely what was actually done. The importance of the evolution of clusters was thus revealed. The Hertzsprung-Russell diagram, described by Henry Russell (1877-1957) in 1913 and independently by the Danish astronomer Ejnar Hertzsprung (1873-1967) around 1905, was plotted. This was one of the greatest accomplishments in astrophysics at the beginning of the century. A point representing a star is plotted in a two-dimensional diagram. The spectral type of the star is indicated on the abscissa, roughly representing a measure of its surface temperature, and the absolute brightness of the star is plotted on the ordinate, one measure of which is its absolute magnitude.

The difficulty of using such a diagram is obvious, however: to ascertain the absolute brightness of a star it is necessary to know its distance! Strictly speaking, the method is therefore applicable only to stars whose distance from the Earth is known.

Under such condition, clusters are of particular interest, since we know that all the stars comprising them are more or less at the same distance from us. If a diagram is then plotted for a cluster, the ordinate is defined to within one constant. But, plotted with apparent magnitudes, the diagram as a whole can shift and is identical to one that would be traced with absolute magnitudes to within one constant.

Now, it is apparent that in this type of diagram, the stars are grouped in "sequences." The position of each star in the sequence is tied to its mass. The most massive stars are lined up among the hottest stars, and the least massive among the coldest.

This empirical result can then be completed through a purely theoretical calculation, by letting a spherical mass of gaseous matter of given chemical composition evolve with time and considering solely that mass, differing from one star to another. By means of a computer and enlisting "digital experimentation," it is thus possible to make a "star" identified with that mass evolve. We find that after a fairly long time the stars have shifted from the initial positions they occupied on the diagram. For a star having a solar mass (the Sun?), the evolution is appreciable in several million years.

This combination of methods makes it possible to date a star cluster. The age of both globular clusters and open clusters can be established very precisely by comparison between the empirical diagram and the theoretical diagram resulting from digital experimentation on a gaseous sphere having a unique chemical composition but a mass arbitrarily selected between one-tenth the solar mass and one hundred solar masses. It has thus been found that globular clusters like the Centaurus cluster are about ten billion years old, while the Pleiades are a thousand times younger: about ten million years old, they are contemporary with the appearance of the first humans on Earth! Other clusters are even younger: so-called "very young" clusters exist, containing stars which have not yet reached adulthood.

One very important fact emerges from these studies: the most massive stars, which, after all, radiate far more than the less massive stars, so to speak "burn away" their hydrogen faster; consequently, they move away faster from the state of an adult star and become old stars much sooner than the others. That is why prematurely "old" stars exist at the age of ten million years; they are stars with a heavy mass. Conversely, one can find "young" stars as old as several billion years, because they are stars with a very light mass.

The stars with the lightest known mass have approximately one-tenth the solar mass. And there are also stars known to have a mass nearly fifty times greater. All in all, that particular group of stars is to be found within a fairly narrow "bracket." This is understandable, when we remember that the energy flow depends considerably on the temperature and that a very slight increase in mass raises the temperature only very slightly but intensifies the energy flow considerably. Under such conditions, the star very soon leaves the range in which it can remain stable.

This, very much simplified, is the description we can offer of the evolution of stars. One type of evolution corresponds to a given mass. Schematically, if the mass is multiplied by ten, the luminosity is multiplied by ten thousand, which means that the lifetime of the star is divided by ten thousand. The Sun has an evolution of its own; a star with twenty or fifty times its mass will have quite a different evolution, one that is far more rapid.

THE END OF THE SUN

What, then, will be the future of the Sun? The Sun will burn away its hydrogen. It is going to burn it away first in its hot central regions where the temperature rises to approximately thirteen million degrees; those central regions will then be depleted of their hydrogen and enriched with helium; on the border of that zone, thermonuclear "combustion" will continue, so that the region where hydrogen is being consumed will gradually expand.

49

But this process will necessarily be accompanied by a change in structure: the temperature of the central regions will increase. As a result, the helium atoms will start being consumed. Nuclear reactions involving the helium will be triggered as the central temperature continues to rise. The star will thus become a red giant, and then a supergiant. What is going to happen next? We do not really know. We can assume, however, that the Sun will remain a supergiant for a time, but that it will steadily lose much of its mass through the "solar wind" that has existed since the most primitive stages of solar life.

The solar wind, at the present time, is reflected by the fact that a small fraction of the Sun's mass is dispersing in space. The rate of this dispersion is very low, since it is only 3×10^{-14} solar mass per year (at that rate, it would take three trillion years to totally dissipate the Sun's mass into space). No solar wind theory is as yet completely satisfactory. Does it emanate from the Sun's interior, or only from the bottom layers of the corona? In any case, that wind truly controls the solar environment. It ceaselessly sweeps up small dust particles and diffuse gases, carrying them across the solar system and beyond. But when the star becomes a red giant its surface will be much greater and its temperature lower; the solar wind will then be much more intense. Previously unobserved phenomena will then be seen to occur very suddenly: the Sun will eject a mass of matter which will become a nebula, a term wrongly referred to as "planetary." At the same time, the remainder of the star will be short of fuel and will collapse. All that will be left is a very small-sized star, known as a "white dwarf." Its mass will be somewhat less than that of the present Sun's, but it will no longer have an energy source other than that of the contraction driving it. Such a prodigiously compressed object will be so dense that ten tons of it would fit into a thimble! Let us add that this object will retain the angular momentum of the Sun and hence turn very rapidly and that, since its surface will be small, it will radiate very little. This white dwarf star may then remain in that state for millions or even billions of years. What is certain is that it will end up being extinguished; but its lifetime will far exceed anything hitherto known.

The truth is that very few white dwarfs are observed.

As they are very small, only those that are very close are accessible to us. We are, for example, familiar with the companion of Sirius, a very characteristic white dwarf, which is the end product of the normal evolution of a star comparable to the Sun.

When will we become a white dwarf, one may ask? We will first have to go through the stage when the Sun becomes a red giant. But at that time we earthlings will experience a catastrophe: the Sun, as we know it, will end. The planet Mercury, and then Venus and Earth will be "enveloped," swallowed up. And even if the red giant thus formed is colder than the present Sun, it will still be hot enough to burn us all to a cinder! The lifetime of a star like the Sun is on the order of some ten billion years. From the time of its formation until adulthood, it has lived 4.62 billion years. It thus has several billion years of life left, so there is clearly no reason to succumb to catastrophism: if Earth is threatened, it is not by the Sun. Let us now consider a star of twenty or fifty times the solar mass. It will consume its central hydrogen; the internal reorganization we have just described with regard to the Sun will be very rapid; helium reactions will then be triggered so suddenly that the center will not have time to cool down before it is totally consumed. Carbon reactions occurring at even higher temperatures will begin, and further temperature increases will ensue. And, before any cooling, reactions involving elements even heavier than carbon will have already begun, and so on in a chain reaction ending with iron. Iron is the most stable atom of the existing series of atoms. It cannot be consumed. But the temperature continues rising, so that when the star's interior becomes a core of iron, of gaseous iron, phenomena of enormous instability occur: the star becomes a "supernova."

That describes the ejection into space of almost the entire mass of the star which, so to speak, completely self-destructs! No more than a small remnant is left at the center, far less massive than the star (in contrast, as has been seen, to the white dwarf, whose mass will be barely less than the Sun's). In the case of a star of fifty solar masses, the remnant will be just a few solar masses. And that remnant is going to collapse much faster, much more violently than the white dwarf. The physical state of that object will therefore be different, in that we shall be dealing with a "neutron star." In such a star, the nuclei are not compressed, as they are in a white dwarf, but are shattered; compression takes place at the lowest rung, namely, the level of the elementary particles constituting the nucleus. This compression is so intense that it is no longer 10 tons, but nearly 10 million tons that fit into a thimble! Such is then the extremely dense, extremely small neutron star, driven at a fantastic speed of rotation. These stars are being observed today: they are "pulsars." They are a source of radio emission and x-ray radiation that are highly directional because of the structure of the star's magnetic field. The theory of electron rotation in the magnetic field tells us that deceleration of these electrons is at the origin of emission of what is called "synchrotron" radiation. As the compression of these stars is very great, their period of rotation is one second and sometimes even much less. A star of the Sun's mass turns on its axis in one second! Observation of these pulsars, which shine at us like rotating beacons, began about twenty years ago. We owe that great discovery to Jocelyn Bell, who was working in 1967 on the radio telescope of the British Nobel laureate astronomer Antony Hewish.

We should not believe, however, that we are dealing with the ultimate state of condensation of matter. That state is represented by the "black hole." The term "black hole" is rather ill chosen, for no "hole" is really involved: what we have is a super-solid, superdense object, so dense that matter cannot escape from it, with light also being trapped. While it is possible to observe pulsars, black holes are detectable only by indirect effects: either through matter which, at the moment it falls "inside," emits highly intense and very noticeable radio radiation and gamma radiation, or through detection of the presence of a mass revolving around a normal star is unobservable other than by means of the apparent motions of the central star.

How are black holes formed? The most prevalent opinion is that they are not the result of a normal process of evolution of a massive star, as I pointed out, to the white dwarf state or to that of a neutron star. Rather, they are explained by invoking the fact that many stars—approximately two-thirds—are double stars. A double star is a small stellar cluster. The lump formed from the initial medium is fragmented into two pieces not necessarily having the same mass. Now, as has been stated, different masses mean different evolutions. In fact, this fragmentation produces two stars, each of which evolves as a function of its mass, so that after a certain time one of them becomes so huge that its matter tends to fall on the other. It is stripped of its outer layers by its neighbor; the evolution of each of the two stars is thus so profoundly disturbed that a black hole can then be expected to occur. As is evident, this is a very special case. It is believed that at least one example of it is known: that of a star around which another star is surely moving, but which cannot be observed directly. That invisible star has a mass on the order of the Sun's. As for its radius, the following reasoning is applied: one never sees eclipses of the larger star by the smaller one since the Earth is almost in the orbit's plane; an upper limit can thus be inferred for that radius, beyond which there would be an eclipse, that is to say a decrease in the brightness of the larger star every time the smaller one passes in front of it. Now, that limit is roughly equal to the theoretical radius of the black hole of corresponding mass. The conclusion is that we are indeed dealing with a black hole.

But the interest of a black hole object lies in the fact that it is not necessarily of stellar nature. In fact, it has been determined that a black hole can exist whose mass is much greater than would correspond to that of a star. Thus, it is fairly likely that in the center of our Galaxy there is a black hole whose mass would (perhaps!) be one hundred million times the solar mass! It must be admitted, however, that we have only very indirect evidence of this.

II

THE SOLAR SYSTEM

AND ITS HISTORY

THE CYCLE OF EVOLUTION
IN THE GALAXY

If the process of evolution of stars is considered as a whole, I believe one point has to be stressed: the entire history of the Galaxy is the history of a perpetual recycling of matter. There is no appreciable disappearance of matter. We have seen that out of the cradle of interstellar matter, nodules or "lumps" are formed, which give birth to stars, since at the end of evolution collapse occurs. But such collapse is accompanied by the ejection from the star of matter which returns to interstellar space where it is recycled. In other words, the diffuse matter in the Galaxy is first transformed inside a star, where it is progressively enriched with heavy elements, especially in its central regions. That enriched material is then dispersed in space, for example when a supernova explodes. This modification of the star's chemical composition makes it impossible for us to compare young and old stars, since they are not formed in interstellar media endowed with the same properties. And interstellar media are progressively enriched with heavy elements.

The system has to be envisaged as a whole, with the interplay of diverse parameters. I have already pointed out that the important factors in the evolution are the series of exchanges taking place between opacity, which retains the energy inside the star, and the production of energy constituted by the thermonuclear reactions in the hot regions. A certain equilibrium is thus maintained. But when the Sun is considered in detail, there are other aspects to its evolution. The Sun, in fact, is not an isolated star; it evolves in a given medium. There is thus interaction between the Sun and that medium. Such interaction is manifested by the fact that radiation is emitted on the Sun's surface and matter escapes. The Sun's surface, a region of maximum interaction between the evolutionary phenomena of the nuclear center and the outer atmosphere, is also the site of most spectacular phenomena. The magnetic field around the sunspots expands into arches, into gigantic loops. And into these arches matter is channeled in brilliant spirals: they are the magnificent solar prominences. The field, manipulated in turn by the movements of matter, is twisted, multiplied, convulsed and condensed. Having become very unstable, it suddenly explodes; the magnetic energy liberated is manifested by a gigantic eruption of light, x-rays, radio waves. Extremely rapid particles, electrons and protons, are ejected into interstellar space. They constitute only part of the stellar wind. Less violently, a massive swell inundates space from all parts of the solar surface, even those that are inactive and barely magnetic. This is the solar wind already mentioned— the solar wind whose dynamics are well described, but whose origin is still little known, and never stops "blowing." It appears to be very irregular and is asymmetrical around the star: in terms of mass of matter ejected, there is apparently far more wind in the equatorial regions than in the polar regions; its velocity also seems much greater in the equatorial regions. The solar wind is a phenomenon which ejects approximately 3.10^{-14} of the solar mass per year at velocities which, when measured in the vicinity of the Earth, are around 500 km/s [300 mi/s], according to data supplied by satellites. It is evident that the whole solar environment—and, therefore, also the planets—is swept by the solar wind. Now, it should be noted that the influence of that wind on climatology is undoubtedly important, and not just for the Earth's climatology, but for that of Jupiter, for example. We are indeed constantly exposed to that fluctuating wind, which is never the same from one point of the stellar environment to another, nor from one moment to another, and which represents a major evolutionary phenomenon for all stars.

Are there then stars whose wind (the "stellar wind") is much stronger than the Sun's? Some are known to lose up to one hundred thousandth of their mass per year. This means that they have a life expectancy of one hundred thousand years, if—although this is not proven—they are destined to lose mass in the future at the same rate as today.

So here is a major evolutionary phenomenon, the solar wind, that is tied to the interaction between the Sun and its environment, which gives concrete expression to its importance.

Astrophysics can therefore no longer be practiced as it was at the turn of the century. It is no longer possible to isolate the planetary system and study individual objects in isolation. Earth, for example, belongs to the Sun. One has to consider it all as a whole in which linkages are boundless: linkages between the different regions of the Sun, between the Sun and the planetary system, between the Sun and our Galaxy, between the different portions of the Galaxy. From these linkages flow the most interesting phenomena.

FORMATION OF THE PLANETS

The Sun is a star. We have seen how stars are born; we now know how the Sun was born. The stars evolve according to different processes as a function of their mass. As a result, conclusions can be drawn as to the future of the Sun. We have just seen that these phenomena must be considered within the free scope of their multiple interactions. We have enough information on hand to explore the question of the planetary system's formation, at least in broad outline.

60

Let me repeat: the planets form an integral part of the Sun. We belong to the Sun. That simple idea, by which astrophysicists are guided today, was established only after a long history.

The question of the origin of planets has been asked since earliest history because the planets, those wandering celestial bodies, were found so intriguing. Furthermore, there was clearly nothing like them in what was otherwise visible in the sky. We therefore had to wait until the 18th century for a coherent theory of the formation of the planetary system to be elaborated. One name stands out: that of Buffon. Georges-Louis Leclerc, Comte de Buffon (1707-1788), propounded, in fact, a *catastrophist* kind of theory, which was admissible as long as the Galaxy had not yet been explored. According to him, the planets were born as a result of a solar accident. In Buffon's view, that accident consisted of the fall of a comet onto the Sun, producing the ejection of masses of matter; when these then cooled and condensed, they were transformed into those huge grains of dust he believed to be our planets! Though he can be regarded as the first, Buffon is not the only one to have propounded such a catastrophic origin. For example, George Darwin (1845-1912), Charles Darwin's son, sought to perfect Buffon's theory. Such explanations had a very strong and persistent appeal. Not so long ago, in fact, eminent scientists like physicist Sir James Jeans (1877-1946) and Alexandre Dauvillier (1892-1979), supported such theories. Specifically, they imagined that the Sun's ejection of masses destined to become planets was associated with the passage of a nearby star; through a tidal effect, the star was believed to have torn away matter from the Sun. Unfortunately, these catastrophist theories have two essential defects. The first is their inability to explain adequately the distribution of angular momenta in the solar system; the second is their inability to account for the dispersion of matter in that system. Why are the planets closest to the Sun those with the smallest mass? That does not fit in at all with such theories. Some *ad hoc* hypotheses can, of course, be adduced to explain things; but should not *ad hoc* hypotheses be avoided as much as possible? That, for example, is what Karl Popper explained in *The Logic of Scientific Discovery* (1959). But earlier Pierre Simon Laplace (1749-1827) had made an observation along the same lines; he was a follower of Newton and the Newton-Descartes controversy had ended in Newton's favor. Laplace thus showed how, in retrospect, the Newtonian theory was refutable, while the Cartesian theory was not. It is always possible, he explained, to adapt the theory of turbulent flux to any situation that might arise; a finishing touch here or there can accommodate matters. In other words, the theory is not "refutable"; it is not a good theory.

In the face of Buffon's theory and that of those who held similar views, Immanuel Kant (1724-1804) and Laplace propounded another explanation of the origin of the planetary system. Even if their theories were not identical, together they advanced a major hypothesis (which would come to be known as the "Kant-Laplace nebular hypothesis"). Although they remained qualitative, these theories, through that hypothesis, had a decisive influence on the course of research. The new idea is as follows: everything is supposed to have begun with cold nebulous matter in condensation; while the center condenses, the rest— not yet condensed—forms lumps smaller than the central lump. From those lumps originate the planets. This hypothesis has the merit of explaining how all the planets turn in the same direction around the Sun; moreover, it becomes possible to understand why their axial rotations are in the same direction as their rotating motions around the Sun.

This hypothesis, at least the way Laplace presented it, still left several points obscure. For example, it does not explain why most of the angular momentum of the solar system is duplicated in the planets, but not in the Sun. This was a formidable problem that the cosmological theories very soon came up against, foremost among them that of the German astronomer Carl Friedrich von Weizsäcker (born in 1912), who imagined systems of eddies in which the appearance of the planets is set in motion. Von Weizsäcker's system was somewhat simplistic as far as distribution of the eddies is concerned. That prompted Soviet scientist O. Schmidt (1891-1956) and Gerard Kuiper, an American astronomer of Dutch origin (1905-1973), to expand upon the theory in order to make it somehow more realistic. Evry Schatzman showed how the speed of rotation of the planets could be calculated numerically in a theory like von Weizsäcker's. The major argument in favor of the nebular theories is, on the one hand, the properties of the planets that they are capable of explaining and, on the other, the fact that it has been possible to observe, among the stars closest to the Sun, at least one other planetary system. The argument can be regarded as decisive, for if Buffon and his successors had been right, there would be very few planetary systems, the initial accident having little chance of recurring. Now, it indeed seems that, among the twenty stars nearest the Sun, the presence of planetary systems can be reasonably surmised around a good third of them.

To understand how this conjecture came about, some familiarity with the techniques making it possible to uncover evidence of such a system is necessary. Direct observation has only very recently been possible, thanks to satellites operating in the infrared range. A very sharp photograph is thus available of the environment of the star Beta in the constellation Pictor ("the Painter"), which can be interpreted as a planetary system. There are, of course, other interpretations that view it as a simple double jet of matter. Vega, a star close to the Sun, bright and markedly hotter than the

Sun, seems to be surrounded by a system radiating in the infrared. Now, if a body radiates in the infrared, that means it must be cold. If it is cold, we must conclude that it is planetary in nature. Vega is thus believed to be at the center of a protoplanetary system.

But other resources are available for uncovering evidence of such a planetary system. It can be detected indirectly: imagine a star surrounded by planets; that star and those planets are going to move around the center of gravity of the system and, consequently, even if you cannot see the planets, you are able to observe the movement of the star. From that movement you can calculate the mass of the companion. If that mass is on the order of the solar mass, one can say that it is an invisible star because it is not as bright and very close or, in exceptional cases, might be a black hole. But if a mass is discovered which has one-twentieth the solar mass, the conclusion is that it cannot radiate. Then it is not a star and can only be a planet!

An interesting object, Barnard's star, is always mentioned in that connection: Peter Van de Kamp (born in 1901) showed that this star could be explained only by the existence of at least two planets. Now, let us suppose an inhabitant of a planet around Sirius was observing our solar system; by these indirect methods he would detect only one planet having a notable mass. He would, in fact, be able to observe only Jupiter directly. The whole solar system would thus be represented by one "average" planet, the mass of which would approximate that of Jupiter. Then, the assumption we must accept, namely the existence of two planets in the case studied by Van de Kamp, is a good indication in favor of the existence of planetary systems. If that indication is confirmed, if one-third of the stars close to the Sun have planetary systems, that will be a very convincing argument in support of the nebular hypotheses.

Actually, there are many other theories which today merit attention. Some introduce the magnetic field in the organization of the protoplanetary nebula. For example, the Swedish physicist Svante Arrhenius (1859-1927) developed a magnetic theory of the formation of stars at the end of the 19th century; so did Fred Hoyle more recently. These are essentially clever variations on the nebular theory which make it possible to explain the distribution of magnetic fields in the planets, as well as other phenomena observed.

NEW MEANS OF OBSERVATION OF THE SUN FROM THE GROUND

An important idea, which originated a few decades ago, completes this exploration. It is the possibility of almost direct probing of the interior regions of the Sun either through the study of seismic wave propagation or of the flux of neutrinos emitted in the central regions of the Sun. The photons (light quanta) are very rapidly absorbed, re-emitted, and reabsorbed, so that the luminous energy produced in the center of the Sun takes hundreds of thousands of years to leave that body. The same is not true of neutrinos. Neutrinos are the most energetic particles produced by the solar machine. Involved here are leptons, that is, elementary particles which do not have a strong interaction and whose mass at rest remains unknown, but is probably very small. The solar matter is practically transparent to them, so that the neutrinos emitted by the central regions of the Sun (like the very high-energy photons emitted by the same thermonuclear reactions) depart those regions, cross the Sun in a matter of seconds, and reach us in eight minutes! The role of these neutrinos proves very important, because the neutrino flux is very sensitive to the nuclear reactions in the center of the Sun. For a long time the results of observation of the neutrino flux have not been very well understood, such observation having begun, after all, only about twenty years ago. That is understandable: almost everything is transparent to them. They are therefore virtually invisible! Neutrinos pass through most instruments. Hence, in order to observe them, it was necessary to develop very elaborate devices, and that took time.

Today we have several neutrino "telescopes" at our disposal. The most famous one is the Davis telescope, installed at the bottom of an old gold mine in

South Dakota at 1,800 m [5,900 ft] below ground level, in order to be protected from cosmic rays emanating from nowhere other than the Sun and which would cause disturbing secondary effects. The receiver is immersed in a water tank to prevent radiation due to the radioactivity of the surrounding rocks from producing artifacts, which might induce a false belief in the existence of neutrinos of solar origin.

There is also a neutrino detector buried deep on the Italian side of Mont Blanc, one in Japan, still others elsewhere and several in the planning stage. Thanks to the experiments set up by Davis, which date back furthest, it appeared that the neutrino flux was three times less than was predictable from the standard theory of the Sun. A factor of three is not to be scorned, even in astrophysics! Scientists then lost themselves in conjecture to account for this unexpected disparity. They wondered about the validity of the neutrino theory; wasn't that theory still very incomplete? But they also again called into question our model of the Sun. Evry Schatzman and André Maeder hypothesized that the regions of the Sun situated immediately around those where the nuclear reactions occur are driven by turbulent motions, which they surmised had the effect of reinjecting unconsumed hydrogen toward the center of the Sun. A change in the rate of production of neutrinos in the center of the Sun was assumed to follow. This hypothesis opened up the possibility of fitting theory to observation. Unfortunately, it seems that this theory, excellent as far as neutrinos are concerned, spoils our solar model in another respect: it does not allow the Sun's oscillations to be satisfactorily explained. There is then every reason to believe that the contradictions will be removed only by progress in the physics of neutrinos or by a modification of the basic data used in modeling of the solar interior.

The phenomenon of oscillations, as weak as they might be locally, namely the velocities it implies, should not be regarded as secondary. On the contrary, it is another wide-open and active field of research. Generally speaking, how can an oscillation be explained? A piece of matter is displaced, and that piece of matter returns to its position of equilibrium. Around that position of equilibrium, it oscillates. What then is the "return" force, the force which brings the piece of matter back to its position of equilibrium? It appears that several kinds of return forces are possible.

First of all, there are "pressure or forced oscillations": if you compress a gas, the return force consists of the pressure which tends to push back the compressed gas. These are also called "acoustic oscillations," because they occur in the same way in organ pipes. As far as the Sun is concerned, they are the oscillations that have been observed most successfully. But there are also gravity oscillations: for example, the swell of the sea, where the return force is weight, without any compression. It cannot be positively asserted, but it seems that it has been possible to detect such oscillations in the Sun. The importance of pursuing the study of gravity oscillations lies in the fact that, in theory, they are believed to provide the greatest source of information on the central regions of the Sun, whereas pressure waves, as they have been observed, tell us something only about the somewhat more external regions. To probe the Sun in depth, observation of both types, constituting the two faces of helioseismology, would be needed.

For example, the study of pressure oscillations has been enlisted to show (though the demonstration was disputed!) that rotation of the interior layers of the Sun is far more rapid than rotation of the outer layers. As the Sun turns, the average speed of rotation on its surface is one revolution in twenty-seven days. That is to be understood as an average, for actually the equator turns faster than the poles. In the center of the Sun, rotation is believed to be almost twice as fast (according to the already mentioned interpretations of the characteristics of acoustic oscillations): one revolution in two weeks. There is thus said to be an actual discontinuity within the star, a discontinuity that would really complicate the machine.

Failure to understand such discontinuity of rotation, if confirmed, is tantamount to not understanding fully the convection, magnetism, and solar activity contingent on it in several respects. That is why studies of pressure and gravity oscillations can be considered very promising, and it is reasonable to devote a great deal of effort to them. They will lead to important knowledge not only of the center of the Sun, but consequently also of the interior of the stars. Thus, a sister star of the Sun is already known, which has practically the same properties, Alpha Centauri: same mass, same brightness, same spectral classification. And yet its oscillations are not the same... Why is that? Some day an answer will have to be found to that type of question.

In our sky, the Sun is infinitely brighter than any star. Indeed, ten million times more light comes to us from the Sun than from all of the rest of the Universe! This offers obvious advantages for observation: instruments enabling us to study its surface details, as a function of time, can be built with an extremely high resolving power, without necessarily requiring very wide apertures or being great light collectors. There is always plenty of sunlight! That is why instruments for observing the Sun have narrow apertures—not telescopes some 15 meters [50 ft] in diameter! With an instrument about 1 meter [3 ft] in diameter, spectacular observations can be made! But it must be added that the secondary instrumentation used for image analysis has to be very elaborate, not to say sophisticated, in order to exploit the potential wealth of information to best advantage.

One often hears the remark: "the Sun is just one of several stars; too much time has already been spent observing it. Let us get on to something else." That seems to me a big mistake. The Sun still remains an exemplary celestial body: the only star on which stellar activity can be studied, on which our theories of the internal structure of stars and our ideas of their evolution can be verified.

Since Galileo, astronomers have been accumulating observations of solar activity; for two centuries after him, these observations have remained admittedly highly qualitative. On consulting old chronicles, some very interesting things turn up. For instance, the observations of Father Scheiner of Harriott (1573-1650) or Philippe de La Hire (1640-1718) or Nicolas de La Caille (1713-1762) are marvelous. Substantive observations, however, are recent; general study of solar magnetic fields dates back less than a century and their detailed study just a few decades. Now, one solar cycle is equivalent in duration to approximately twenty-two years (twice eleven years, since the magnetic polarity changes direction from one cycle to the next). Maximum solar activity is thus generated every eleven years. But there are also longer cycles that we still know very little about: of seventy-five to eighty years, but undoubtedly also of two hundred or three hundred years, which we have not yet been able to observe directly!

There is thus a great deal to be learned from observation of the Sun; and what are sometimes called "routine" observations, using the occasional pejorative connotation of the term, can actually be very fruitful. Watch regularly: for days nothing happens. Then suddenly an unexpected and often fabulous phenomenon occurs, which must not be missed.

One finds the attitude of numerous astronomers and often of the authorities distressing. A remarkable instrument, unique in the world, used to be available in Sydney, Australia. It was a radio-heliograph, laid out in a ring two kilometers [a mile and a quarter] in diameter, consisting of thirty-six parabolic antennas, all trained on the Sun and following it over the course of time. That gigantic instrument yielded images of the Sun at several wavelengths and made it possible to monitor its activity regularly within the range of radio waves; essentially, it took advantage of the radiation emanating from the solar corona. But now it has been shut down. At Nançay, in the Sologne region of France, an instrument whose performance is comparable to that of the Sydney installation, has been constructed; but when the Sun sets in Nançay, it rises in Colorado. However, there is no solar radio telescope in Colorado. And when the Sun sets in Colorado, it rises in Australia, where there no longer is a solar radio telescope. Result: solar activity today is no longer being regularly monitored in the radio range; nor is it even being followed very well in the range of ordinary light ("optical" range).

How did such a situation come about? In the Australian case, a combination of inadvertent circumstances arose: the radio astronomer who had organized a team around him was appointed to one of the highest administrative posts in Australian research, another took his retirement, a third died, and so on.

The fact remains that, despite the resolutions adopted by every committee of solar physicists, no cooperative solar research network currently exists, notwithstanding the unanimous desire for one expressed by the physicists concerned. This situation is attributable in large part to the budgetary competition being waged in the different areas of astronomical research. Thus, when people in France want to build a solar telescope, like the wonderful Themis project planned for the Canary Islands, the decision is difficult to make because they also want to participate in the great interferometer telescope for the European Southern Observatory at La Silla Mountain in Chile, not to mention numerous other ground or space projects. And in this competition the solar astronomers are not in a very good position because, as I stated, the opinion is widespread among astronomers, albeit wrongly so, that the Sun is sufficiently well known; solar observations seem to them to be lacking in attraction and exoticism.

Yet, we are dealing with basic astronomy because the Sun is a veritable laboratory. We have seen how this is so in connection with the neutrinos, with "neutrino astronomy"; but the Sun also lends itself perfectly well to the study of energy conversions of all kinds—radiation transfer, dissipation of mechanical or hydromagnetic waves, eruptive phenomena, etc.

And since I am determined to fight for continuous observation of the Sun, I can do no less than champion the cause of the very symbol of an observatory, the Pic-du-Midi Observatory, where Lyot (in 1930) installed the first coronagraph. Also at the Pic Observatory Rösch, Macris, Muller and Roudier have since then amassed the world's best photographs of solar granulation, which have contributed essential knowledge of the physics of convective motions— and will continue to garner new information. And I am not even referring to the lunar or Martian photographs, nor to the study—ongoing, year after year— of Venus (Boyer and Camichel) or of Mars (Dollfus and his successors), or of Jupiter (Focas). The Pic Observatory is now threatened (it is a question of "saving" it, we are told, for its operation requires a lot of money) by a large-scale tourist and cultural complex. There is no small risk that the exceptional qualities of observation at the Pic will deteriorate: another example of misguided zeal! Why is that so? Because hard choices, it seems, have to be made in budgetary allocations.... Would it not be preferable rather to *separate* the financing of the necessary large-scale, highly expensive and distant international operations from the funding of regional (for example, European) operations, which are also on a large scale and expensive and often likewise distant? To separate these, in turn, from national operations (in France, the medium-sized observatories affording the opportunity for specific projects and especially regular long-term monitoring of variable stars, the Sun, planets and stars—like the Pic-du-Midi or Saint-Michel observatories) and from university operations (teaching observatories, hence close to a university)? Should the Pic Observatory be measured against Themis? On the contrary, their complementary nature is obvious: one should remain French and Pyrenean; the other is to become European. And why not, instead of linking the Pic with the University of Toulouse, envisage a new, small but quality university in the Department of Haute-Pyrénées, at Bagnères or at Campistrous, with an American-style experimental campus and a limited but resident faculty? It is not a dream! The Pic would be the jewel in the crown of that university. And a jewel where good work must be pursued, and where good observers must be trained.

ROLE OF SPACE RESEARCH

The Sun offers us opportunities for observation, if not for experimentation, of physical phenomena whose parameters we can vary, as it were, with pressure, temperature, and density conditions and dimensions that we cannot match on Earth. It is not only an astrophysics laboratory, but also a physics laboratory. Is that not the best justification for detailed and continuous observation of the Sun? It is undeniable that the operation of such a global or world network is costly: solar instruments are perhaps smaller than others, but they are also very delicate and demand rather heavy investments. For example, just think of the measurement of magnetic fields on the Sun: image processing is a highly sophisticated operation. But the results are magnificent: maps replete with information are produced, on which one can see the local evolution of the fields and the manner in which certain structures are destroyed because they are too complicated, so that the stored magnetic energy is released and produces a solar eruption. What we see there cannot be seen any other way, nor on the other stars; but it does require highly sophisticated instruments. The solar telescope that France is to construct in the Canary Islands ("Themis") is intended for magnetic studies, not only of the Sun as a whole, but of the details of its surface and its different regions.

What about space research in our field of solar physics? There is no doubt that it has contributed a great deal, since it has made it possible, in particular, to observe radiations that could not be previously studied: ultraviolet radiations, x-ray radiations, and infrared radiations of intermediate wavelengths between ten microns and one millimeter. None of these radiations was previously accessible to us; what had been available, so to speak, were only one octave of solar radiation in the visible range and three octaves in the radio range. Now, the possibilities have been greatly expanded: the entire spectrum is covered. It can be said that with radio astronomy space research has in the last thirty years helped revolutionize our knowledge of the Sun.

One serious problem remains, however. Solar instruments must be highly reliable and, as just stated, they are very delicate. To date, the few instruments and the few solar satellites which have been placed in orbit have had only a brief life span; the gain in reliability is at the expense of extreme fragility. Thus, for quite a long time there was no solar instrument in orbit that could observe the Sun in the x-ray waves. Agencies like NASA, when asked for x-ray images of the solar corona—those famous splendid images showing the coronal holes stretching out and the hot arches glowing—always supply the same ones which were taken during missions dating back quite a few years. Yet, observation of the corona, aside from its aesthetic attraction, is of great scientific value: we have seen that the solar surface is the layer of the Sun where opacity suddenly increases. Underneath, the medium is opaque; it is hidden from us. But outside of this actual "surface," there is still a mass of matter, transparent in the visible range and driven by the solar wind. It is that mass which constitutes the solar corona.

The corona is highly visible during total eclipses of the Sun. In the absence of an eclipse, its essential characteristics can be measured by means of instruments called "coronagraphs," invented by a French optician of genius, Bernard Lyot (1897-1952) in the 1930s.

The solar corona is a mass of very hot gas: mechanical and hydromagnetic waves propagated from deep regions are dissipated there. As the medium of the corona is very dilute, it is readily heated by these dissipations, so that the temperature of the corona reaches an average of one million degrees. It must be added that, while the corona is transparent to visible radiation, or light, it is opaque to x-rays as well as to radiometric radiation: at these wavelengths, the corona is seen "in front of" the solar disk. Furthermore, it is apparently not spherical. In contrast to the zones of the Sun's atmosphere called photosphere or chromosphere, which lie close to the visible surface, the corona is so highly controlled by the Sun's magnetic structure that the streamers of matter for which it is the stage, originating from combinations of magnetic forces and mechanical forces, have extraordinary shapes and are devoid of any spherical symmetry. Gravitational forces are no longer involved. More precisely, at the distance from the Sun where the coronal medium is present, the gravitational force is secondary to the magnetic forces and to the dynamics of the solar wind.

THE FUTURE OF THE SUN

In general, as we know, what makes a star (or celestial body) spherical is gravitational force. As soon as that force is no longer dominant, there is no longer any reason for the object to be spherical. Hence, the magnificent display presented by the corona. For a long time it was believed not to be a solar phenomenon at all, but rather exhalations from the Moon revealed from time to time during total eclipses of the Sun. It was believed that the famous "streamers" belonged to the Moon's atmosphere, which was thought to be far more irregular than that of the Sun. By the end of the 19th century it became evident that the corona is really solar, that its shape is molded by the action of the wind and magnetism, and that as far as the Moon was concerned it is too small to retain the slightest vestige of an atmosphere above its mineral soil.

Now, let me repeat, never has observation of the Sun in the x-ray waves yet been undertaken on a regular and continuous basis. The same can be said, moreover, about the planets.

FOR CONTINUOUS OBSERVATION OF THE SUN

The present situation is not healthy. Paradoxically, technological progress is turning against the very interest of research. Indeed, owing to the fact that many things previously inaccessible to us can now be observed, there is a tendency to think that ground observation is no longer necessary. Thus, solar research is relying upon satellites which, in fact, observe at wavelengths hitherto unknown to us; as for the planets, we are applying ourselves to sending probes there for on-site viewing. But the reverse side of the coin is that, little by little, we are tending to replace admittedly mundane but continuous ground observation of the Sun and planets by very wonderful and very precise observations, to be sure, but covering unduly short periods. A fascinating type of research is thus being missed: to mention one example, the climatological evolution of Jupiter is no longer being studied; to mention another, as I pointed out, the Pic-du-Midi Observatory, which obtained the best photographs of Mars, is threatened with tourist exploitation, if not with actual closing, even as NASA is going to use those photographs for the planned landings. It is known that the surface of Mars is changing constantly, that the winds there are awesome and sweep the planet from north to south, shifting enormous masses of sand. It would therefore be very useful to be able to follow the evolution of the planet continuously. Now, such a study can be made only from the ground, and not from space.

It should not be concluded from these remarks, however, that space observation is useless, let alone harmful. There is, indeed, information that could not be obtained by any other means: it is believed that, for example, the distribution of luminosity on the Sun's surface is not believed to be isotropic and that the polar regions of the Sun do not radiate like the equatorial regions; but as we are situated almost on the Sun's equatorial plane, we see the poles only obliquely, never face on. They escape our observation because of the effect of perspective. A rocket probe leaving the ecliptic and observing the Sun at its pole would thus be an extraordinary means of supplementing our knowledge, provided it could remain there long enough and return several times to establish whether, in the course of the solar cycle, the difference in radiation between the equator and the poles varies. With such observations, we could conceivably make sense of that possible variability. That would enrich our store of knowledge not only of the poles, but also of the sunspots which are found to be always located at low latitudes on the solar disk, and which, it has also been found, without yet any reasonable explanation, do not occur in the neighborhood of the poles.

And, of course, so would our knowledge of the workings of the solar cycle be expanded. The solar probes sent into interplanetary space now prove to be a far more precise and reliable means of study of the solar wind than traditional observation of the tails of comets. The composition of that solar wind (electrons, protons, and some heavy particles) could thus be determined *in situ*; it was observed that the ratio of the number of helium nuclei to the number of protons varies with wind velocity. And thanks to these measurements, it was possible to determine, as we have seen, the annual loss of matter undergone by the Sun.

CYCLE AND MIGRATIONS OF SOLAR ACTIVITY

One of the questions that still remain open today to the theoretician of the Sun is that of the solar cycle: even after centuries of at least partial observation, and notwithstanding the numerous calculations that have been made, the workings of this cycle are still not very well understood. It is from the observation of sunspots that the idea of such a cycle was conceived. Ever since Galileo's observations, the number of sunspots has been known month after month, day after day. As early as 1873, the German pharmacist and astronomer Samuel Heinrich Schwabe (1789-1875) had noticed a certain periodicity in the variation of the numbers of spots. A peak is known to be reached every eleven years. We have thus sought to study the laws of that cycle. But it was soon realized that the number of spots is not the only important characteristic quality of the cycle. That number varies from one cycle to the next, and from the southern hemisphere to the northern hemisphere. Not only does the number of spots vary, but so does the region in which they appear. Starting at high latitude (approximately 45°), that region approaches the equator in a slow migration. As certain phenomena (bright points) also appear at the highest latitudes, before emergence of the spots at about 45° latitude, it is believed that a general migration of activity from the poles to the equator is involved. A migration lasts from eighteen to twenty-two years, followed, at an interval of eleven years, by another migration.

From one "cycle" to the next, from one "migration" to the next, the magnetic polarity of the phenomena changes. In a period of minimum activity, the Sun's (heliographic) north pole is also (we assume) a positive magnetic pole; the south pole, a negative pole. The generally double spots which then appear have a double polarity: in the course of a given cycle, there is a positive pole for the head spot, the one preceding the other in rotation, and a negative pole for the tail spot in the northern solar hemisphere; in the southern solar hemisphere the polarity of these minimagnets is reversed and the head spot has a negative polarity. In the following cycle, the phenomenon is reversed and the polarities are replaced by those of opposite sign.

Around the localized magnetic phenomenon of a group of spots, the dazzling and majestic epiphenomena of solar activity, prominences, and eruptions, develop in every layer of the chromosphere and corona; they stamp their effect on the distribution of density and velocity of the solar wind. The discipline that delves into the question of cycles is called "magnetohydrodynamics," or MHD: it combines Maxwell's electromagnetism and the dynamics of continuous and compressible media, as defined by the mechanics of continuous media. Since the magnetic energy localized in the solar medium is of an order of magnitude comparable to that of mechanical energy, that medium is the ideal area of application of MHD. But it is a difficult discipline: the equations are not linear and, therefore, have to be linearized, with the result that, except for a few specific problems, the combined solutions of the equations of hydrodynamics and electromagnetism still remain faltering. The largest computers enable us to write these equations, but not to solve them!

Furthermore, the geometry that has to be applied is extremely complicated: not only is it a geometry of spherical symmetry, but the effects of axial rotation and, as has been seen, of differential rotation have to be added to the spherical symmetry (which is not even present in the corona). This type of geometry entails a very large number of calculation network links; it is therefore necessary to resort to very powerful computers in order to arrive at solutions.

However necessary data processing may be, I maintain that in the field of solar physics observation must remain paramount, observation that is continuous, second after second, minute after minute, day after day, year after year. The important thing is also to establish a constant exchange between the observer, who amasses well-chosen and good-quality data, and the theoretician, who tries to understand them without oversimplifying the equations.

III

ASTRONOMY,

RESEARCH,

EDUCATION, PLEASURE

RESEARCH DRIVEN
BY CURIOSITY

Why do people engage in astronomical observation? Out of curiosity, out of simple curiosity! Thus, when they study oscillations, discovered almost by chance, it is in an attempt to understand the processes developing in the center of the Sun. Does research really have to be justified by interests other than the advancement of knowledge? It has been argued, for example, that the construction of solar laboratories is justified by the fact that solar energy will be the energy of the future. But it is totally unnecessary to know about solar physics in order to use solar energy! Since that term is commonly used in research circles, let me say that it is not possible for solar research to be "driven" by economic demands associated with the use of solar energy. We should have the courage to admit it in these somber times when economic profitability passes for the supreme value in human life: solar physicists, judged by that yardstick, are at first glance regarded as irrelevant (others, too, moreover! Long live vacation clubs, advertising and bureaucracy!). Astronomical research is generally driven only by the astronomical questions raised. Only by engaging in a "fishing expedition" type of research has it been possible to discover a number of things in this field that could not have been foreseen and to shed light on a large number of phenomena whose importance went unsuspected. Thus it was with the discovery of the solar corona, because there were eclipses. Let us suppose that the Moon were smaller in size or further away from us; in that case, we would never have seen an eclipse of the Sun, and we would perhaps also be ignorant of that veritable gas pump which the corona is, between Sun and wind. There is no doubt, let me stress, that the idea of the transformation of hydrogen into helium has contributed considerably to the progress of research into fusion energy and thermonuclear reactions in general. Knowledge of the solar corona in turn has undeniably contributed to the progress of research in magnetohydrodynamics. Just think of the work of the Swedish astrophysicist Hannes Alfvén, 1970 Nobel laureate in physics, who pioneered magnetohydrodynamics (in 1943), inspired by questions relating to the equilibrium of sunspots. This discipline was then exploited in the study of hot plasmas and in the development of plasma machines with a view to the production of fusion energy.

It is evident that "fortuitous" research, conducted for its own sake, in order to expand knowledge, can have multiple effects that are by nature unpredictable and, therefore, unprogrammable in other fields. Another example is very much to the point: knowledge of solar activity is of considerable importance in climatology; it is even believed to be capable of forecasting weather phenomena on a small scale.

As early as 1930, the Serbian scientist Milanković; had calculated the effects associated with the mechanical properties of the solar system on climate. He had gone quite far in his calculations. But scientists have since done better. Their calculations cover millennia, even millions of years, and make it possible to plot maps of the illumination of Earth as a function of time and of each point on its surface.

For my part, I am convinced, in spite of the contrary opinions of some experts, that meteorology can in the future also derive a great deal of benefit from research on solar activity. Take the case of rainfall. It has been possible to establish that in the arctic regions rainfall is closely correlated with solar activity, just as are the aurora borealis, ionospheric disturbances, and geomagnetism. That is easy to explain: it is indeed in the vicinity of Earth's magnetic poles that the electrons coming from the Sun and carried by the solar wind—hence energy-laden enough to have crossed Earth's magnetic shield—wind around the magnetic field lines of force and produce the aurora borealis. But we still do not understand how that phenomenon is translated into an appreciable increase in rainfall on the ground during periods of solar activity. It has likewise been observed that during those same periods geomagnetism is very substantially intensified. If Earth is actually a magnet, a sort of magnetized iron ball whose magnetism dates back billions of years, geomagnetism is due also in part to the existence of electric currents traversing the ionosphere. Now, the ultraviolet radiation and x-rays from the Sun, emanating from its active regions, act on the ionosphere and, therefore, on terrestrial magnetic phenomena; but how can geomagnetism, associated with localized phenomena in the upper atmosphere, influence physical conditions in proximity to the ground? That is a real question mark.

Here then is an area of research, whose economic and human importance is considerable, but in which almost no account has been taken of solar physics; yet, quite clearly, it is going to have to be taken into account. The knowledge accumulated out of mere curiosity is suddenly going to assume undeniable practical value. We would not be at the point we are if such "routine" astronomical as well as geophysical research had not been pursued for decades, research that some short-sighted administrator might have described as pointless!

Actually, the problem of research motivation, curiously enough, goes beyond study of the Sun, even beyond astronomy. Splitting up the question raised, one might first ask: "Why does a given researcher (I, for example!) interest himself primarily in the Sun and ignore a whole enormous side of astronomy, concerned with the stars, the galaxies, and even the Universe?" To that question there are as many answers as researchers, and I shall get back to it. The second question, on the other hand, calls for what I may be allowed to describe as a solemn warning. It is simple: "Why engage in (pointless) research and not let one's efforts be guided solely by their applications to human life?" This is truly a social problem, and no genuine researcher can answer this question without anguish.

For let us look at today's world. One billion people are starving and billions of others are living wretchedly. The planet is threatened; the living conditions of people still alive are growing worse rather than better; or rather, they are growing worse in terms of the essentials (potable water, breathable air, tolerable climatological conditions) and they are growing "better" in respects that seem to me quite secondary (cable television, high-speed trains, nuclear defense). The only real progress is perhaps that achieved in the field of health: medical researchers are creating miracles; life expectancy is being lengthened. But does not this progress also raise societal problems by prolonging lives of misery and contributing to demographic imbalances? To the ills we have mentioned we must add the exponential increase in the Earth's population, a major source of pollution! And, we might well add, this general increase affects far more (3 or 4 times more!) the poor countries than the rich. So, the risk of confrontations, the threat of massive migrations, desired or not, and the resurgence of nationalism and fundamentalism, combined with insidious conflicts, are causing people to be preoccupied with a lot of things other than simply learning about the world which surrounds—and frightens—them.

Then, they ask—these people of the age of fear— why this pure research, why so much money spent to learn about the volcanism of Io, Jupiter's satellite, or to understand the mechanisms of solar eruptions? Why shouldn't our efforts be guided solely by the conscious, organized and well-planned determination to solve specific problems? For example, desalinating sea water, reconstituting the ozone layer, limiting the demographic madness, etc., are not these subjects to which research efforts should be dedicated, barring others? If worst comes to worst, would it not be better to allocate funds to improving the environment, human health, etc.?

These people of fear, far from understanding the needs of research (astronomical or other), perhaps even displaying a total lack of understanding, tend more and more each day to cling to illusions, to resort to ridicu- lous but tempting remedies: sects and fundamentalism, nationalism and parochialism, but also astrology, homeopathy, and other nostrums—remedies, illusions!

The truth is that disinterested research, research motivated only by curiosity, is deep down one of the noblest expressions of what makes humans unique. The truth is that, whether or not motivated by a specific goal, research often develops unpredictably; its applications may be immediate (consider, for example, superconductivity) or more remote (could Galileo or Newton have foreseen space research and astronautics?). Some research may or may not have any applications. These may be tremendous or minuscule. A paltry grant made to a researcher in the field of mathematical group theory today can, within ten years, or within two centuries, lead to applications entailing investments running into the trillions of dollars.

The truth is that, today more than ever, it is our duty as researchers to proclaim loud and clear the true nature of research and its genuine motivation. That is what it will take to protect our fellow citizens (may I be forgiven for not totally abandoning my optimism!) from the temptation of false remedies and our human society from the indiscriminate disruptions threatening it.

WHY THE SUN AND NOT, FOR EXAMPLE, THE BIG BANG?

But I had raised another question.

Why choose, in a discipline as broad as astronomy, that one star which is the Sun? And can solar research enlighten us about the Universe as a whole?

Granted that study of the Sun is really the study of a rather commonplace star. A good understanding of solar physics opens the way to stellar physics. The Sun's evolution is one particular case of stellar evolution. And stellar evolution is the key to the evolution of our Galaxy—and thus also, in some measure, of the other galaxies. And thinking about the evolution of the galaxies very soon leads to pondering the evolution of the Universe.

It is true that our general postulate is the oneness, "the universality of the laws of physics." And it is true that this "universality" reflects, not the tentative simplicity of those laws, but rather the existence of multiple linkages in the Universe. The question then becomes: in studying the Sun, our Galaxy or the other galaxies, can we detect the presence of the evolution of the Universe, or not?

Here, we should briefly mention the "big bang" and the expansion of the Universe. A number of observations, interpreted within the framework of the physics of continuous relativistic media, seem to indicate that, say, 20 billion years ago, perhaps more (but the numerical value is not important here), the Universe, considerably condensed and hot, suddenly almost ceased to be in balance and virtually exploded. That unique point in the model of the Universe, that universal and gigantic explosion, is the big bang.

After that, in a medium of radiations and of energy in the pure state, in the course of rapid cooling over a few hundred thousand years, the first atoms, the first stars, the first galaxies very soon appeared. That sudden "primordial" cooling froze the Universe, the galaxies and the first stars that appeared in the Universe, in a defined chemical composition—90% hydrogen, 10% helium, and practically nothing else. Further evolution within the stars altered that somewhat: the stars were formed in a medium arising out of that further evolution. The "young" stars then have a chemical composition enriched with heavy elements, while the "old" stars have a chemical composition close to the primordial composition.

Furthermore, the Universe is not necessarily, on the whole, identical to what it was at the time of the appearance of the first galaxies. Today, the medium is more diluted, and the galaxies which form in it do not have the same composition or the same properties as those formed 20 billion years ago.

What does the Sun tell us in this respect? Nothing or almost nothing; the linkages, if any, are too loose, and our theoretical ideas too schematic, even if they can be supported, really to prove anything at all about the "primeval" Universe by extrapolation from the Sun. The abundance of helium and heavier elements in the Sun conforms to this scheme: the Sun, formed, as it were, this morning (4.6 billion years ago is this morning on the scale of universal evolution!), reflects the chemical composition of an evolved gas. But it tells us nothing about how this gas evolved. Thus, galaxies of pure hydrogen (with 1% helium, for example) could become enriched with helium without the need for nucleosynthesis at the time of the big bang. None of the "evidence" of the evolution of the Universe is very solid. It could just as well have always existed; galaxies might have been forming in it incessantly and evolving there incessantly to the state of extinct galaxies; expansion itself, on which the structure is built, could be the result of no more than a physical phenomenon still unknown, such as, for example, interaction with the "vacuum" (or, rather, with space) of photons (light quanta) to which a zero mass is generally assigned, although a nonzero mass could well have such effects (and that is personally what I *believe*—not what I *know*! I *know* nothing!).

In short, while in other respects the big bang and the expanding Universe are enticing and even convincing views of the evolving Universe, they nonetheless remain just reasonable extrapolations and nothing more. And, in any case, although the Sun (in its chemical composition, rotation and magnetism) has retained traces of the archeological evolution which affected the medium in which it was formed, solar physics cannot tell us anything, either for or against, the standard view of the Universe, that of expansion following the great explosion of the big bang.

TEACHING ASTRONOMY

I alluded above to the societal problem represented by the dangers that today burden all science,
due to worldwide growth itself, with its problems of
survival facing humankind, and to the often irrational
manner in which human beings respond to them,
either by rejecting all science or by demanding that
science solve our daily practical problems.

Under these conditions, should astronomy be
taught at all? And how should it be taught? Formerly,
"cosmography" used to be taught in France for the
secondary school certificate, with mathematics teachers in charge. Although they did not cast it in a bad
light, they did, however, present a rather depressing
view of astronomy: one learned the laws of Kepler or
Newton in abstract terms. That course subject came
to naught, especially because of incompetence on the
part of the teachers, mathematicians who had never
taken any astronomy courses. Another area remained
in which astronomy was mentioned: geography. Here
the talk was about meridians, seasons, day, night, etc.
In short, the rudiments of positional astronomy were
presented: the study of phenomena associated with
the relative motions of the Moon, the Sun, the Earth,
and so on. As for astrophysics, namely the study of
the physical properties of celestial bodies, it has
never been established in the secondary education

curriculum; it is approached only obliquely through some physics courses.

That does not seem at all satisfactory to me: the Universe is an entity; how useful it would be to present it as such to students, when today the only knowledge they might gain about it is acquired from scraps of information picked up by way of solving certain physics problems? How sad it is to see the subject of sunspots, for instance, presented only in a lecture on magnetism, as a simple example of general physical laws, and to see the movement of the planets around the Sun treated only in a subsequent lecture on gravity. No coherent view of the Universe can come out of such crumbs of knowledge.

Should a specific course in astronomy then be established in secondary schools? In light of some surveys, this would not seem to be in vain: approximately one-third of the population of France, according to recent studies, believe that it is the Sun that revolves around the Earth, as in the days before Copernicus! But it seems to me that such a course should be totally elective, not mandated: astronomy is not a tool for everyday life, like mathematics, and the study of languages. Like the Earth sciences, astronomy contributes to a knowledge of the world and affords the opportunity to share a very rich intellectual pleasure. In that respect, one can say that it has an educational value; but that value will be all the greater as pupils have free access to it. Why ask children to pass examinations on what interests them? Now, the environment in which they live, and astronomical phenomena in particular, fascinate them. Children ask highly inspired and very spontaneous questions; they often bring up real posers that have sometimes been forgotten.

Under these conditions, should there be a syllabus? First of all, we must respond to the children's questions, often born of their own sky-gazing and sometimes associated with current events. The right procedure would be not just to describe the facts and present them without explanation, but simply to make it clear how a given conclusion was reached. The history of methods, like the history of ideas, would be an excellent guide. The sky, which appears to us as a spherical surface above our heads, actually has depth, some celestial bodies being closer to us than others. The measurement of distances (Earth's radius, distance of the Sun, Moon, planets and stars) is based on simple geometric methods, just like the proof of the Earth's roundness. First, a geometric description, and then a physical one: why are some stars red, and others blue? Why does the Sun radiate? And so on. The essential thing would be to enable students to understand, and not just to lecture them! And we should also use those wonderful tools at that level, the planetariums.

In higher education the problem obviously takes on very different dimensions: there it becomes a matter of training not only teachers but also astronomers and space scientists. It is therefore necessary to start early, by imparting clear ideas to the students in the first two years, even if it should be impossible, considering the lack of professional openings, to commit them to a very early specialization. At this stage, a comprehensive idea of the Universe, detailed and general geometry, miscellaneous objects, stars, planets, galaxies, evolution of celestial bodies, solar physics, etc., should in my opinion endeavor to offer students a coherent, logical and comprehensible view of all of contemporary astronomy and astrophysics. In later stages (notably, in graduate school), it is in association with plasma physics, for example, that astrophysics would find its proper place: there will be students who will then devote themselves to basic research and take an interest in modeling of the Sun, while others may turn toward industry to work on the design of a spacecraft or, more prosaically, on the manufacture of a plasma device intended for a given laboratory experiment, conceived with a view to the industrial production of fusion energy. From the same perspective, astrophysics can be coupled with turbulence theory in order to derive similar benefits.

If I may draw on my personal experience, university teaching remains very disappointing, because all too often it is not associated closely enough with research: in any case, such teaching did not give direction to my own research. On the other hand, at the Collège de France, where each subject is taught at the cutting edge of present knowledge, the structuring of my courses was a great incentive to my work (if not to that of my students): I realized that given fragments of my lectures, like some of my partial reasoning, were questionable. That led me to think more in depth and, on several occasions, to do research in order to fill the gaps in my teaching; for example, in the field of stellar dust, it was in connection with one of my courses that I realized that the electric charge of interstellar dust was wrongly evaluated in the usual calculations.

THE JOY OF BEING AN ASTRONOMER

I cannot end these pages without mentioning the almost physical emotion which certain astronomical phenomena can instill in you. That would be reason enough, it seems to me, to justify all the efforts devoted to study of the Sun and to astronomical research in general.

I shall never forget the eclipse of the Sun on January 25, 1952, in Khartoum, Sudan. That city is at latitude 15° north approximately. The North Star could then be seen at the same time as the Southern Cross! Simultaneously, there was the Sun eclipsed by the Moon, the magnificent solar corona, Mars, Venus, the Magellanic Clouds and the Milky Way. And, in the distance, an illuminated horizon surrounded the patch of shadow where our camp stood for a few minutes. A spectacle unique and sublime!

But, more modestly, look at the Moon through a small telescope and let yourself be spellbound by the sight of the mountains you discover there. Seeing, then understanding, and understanding a little better still, heightens that already intense pleasure. Every human being is sensitive to it.

One last recollection to illustrate what I have just said. I was at the Pic-du-Midi Observatory one day with an old Greek friend, the late Jean Focas. He beckoned and asked me to come discover an extraordinary spectacle in the telescope. I looked and saw nothing but a pale fuzzy disk. He guided me, taught me to look, to discern details. That went on for an hour. And, suddenly, I saw Mars coming out of nowhere for me, Mars with its "fountain of youth" and its yellow sands. That joy was not associated with any scientific knowledge, but with that of seeing, of finally knowing how to see!

From that elementary pleasure, it is possible, by dint of hard work, to pass on to another, more "scholarly," more intellectual pleasure, if you will, but no less intense. I once pondered the following problem: the layers of the Sun one sees give the impression of being arranged in strict concentric spheres. Therefore, the ray of light reaching us from the Sun's extreme edge must be tangential to the sphere. In reality, though, for many reasons, it is actually a sphere and the ray coming from the edge is not tangential; it is inclined in relation to the layers from which it originates. Now, the observation made from Earth is not superficial; it penetrates the solar layers in depth. Hence, an error committed in construction of the model, if it is assumed that those layers are spherical: the nontangential ray coming from the edge informs us about layers less superficial than if it were tangential to the surface. The deceptive effect was discovered in the 1940s by Roderick O. Redman; he called it a "roughness factor" and I personally greatly enlarged on that question with him, within the context of my thesis. I then had occasion to return to the subject on several occasions, as the observations were progressing. Two years ago I applied that idea (with S. Dumont and Z. Mouradian) to the analysis of space observations made of the intermediate regions between the chromosphere and the corona, that is, the regions where the temperature varies from ten thousand to half a million degrees Kelvin and where rising temperature layers are observed, which fit into each other rather like fingers into a glove. The geometry of those layers proves very complex. I realized then that the "roughness factor" made it possible to interpret the observations much better than had been done previously.

The calculations, curves, etc., all looked great. And that, once again, was an unequaled pleasure: the joy of grand spectacles; the joy of understanding how they unfold, of analyzing their workings.

For the pleasure of understanding phenomena equals that of actually seeing them.

One might think this is a selfish and elitist pleasure of the professional astronomer. But that would be a grave mistake. First of all, because, as I have said, a pair of good binoculars uncovers for the viewer magnificent celestial landscapes, mountains of the Moon or gaseous nebulae. But also insofar as photographs (slides) of the beauties of the heavens have become accessible to us for the past few years at modest prices, as well as films, and books (often superbly illustrated!). It is well known that little room has been made for science on the television screens, at least in some countries; however, it is astrophysics which perhaps is least undermined by that situation, owing undoubtedly to the exceptional beauty, recognized by all, of the images that today's telescopes offer the amateur astronomer of the eternal attractions of the sky.

BIBLIOGRAPHY

L'Astronomie, special issue on solar activity, 1980, Société astronomique de France.

BERTHOMIEU, G., and CRIBIER, M., pub. *Inside the Sun*, proc., 121st I.A.U. series, Ed. Kluwer, 1990.

COX, A.N., LIVINGSTON, W., and MATTHEWS, M.S., (symposium published by), *Atmosphere and Internal Structure of the Sun*, Univ. of Arizona, Tucson.

DELSEMME, A., PECKER, J.-C., and REEVES, H., *Pour comprendre l'Univers* [Understanding the Universe*], De Boeck-Wesmael, university ed., 1988, "Champs" series, Flammarion, 1990, repub., in press.

Les Etoiles, le système solaire [The stars, the solar system*], in *Encyclopédie du Bureau des Longitudes*, last edition 1986, Gauthiers-Villars.

GIOVANELLI, R., *Secrets of the Sun*, Cambridge University Press, 1984.

HOYLE, F., *Astronomy and Cosmology: A Modern Course*, San Francisco, Freeman & Co., 1975.

JORDAN, S., *The Sun as a Star*, pub. NASA-CNRS, 1981.

MILLER, R., and HARTMANN, W.K., *The Grand Tour: A Traveler's Guide to the Solar System*, Workman Pub. Co., 1980.

MONTMERLE, T., and PRANTZOS, N., *Soleils éclatés* [Exploded suns*], pub. CNRS-CEA, 1989.

NOYES, R.W., *The Sun, Our Star*, Cambridge, Harvard University Press, 1984.

PECKER, J.-C., *Sous l'étoile Soleil* [Under the Sun star*], Fayard, 1984.

SCHATZMAN, E., *Our Expanding Universe*, "The McGraw-Hill Horizons of Science Series," McGraw-Hill, 1992.

SCHNEIDER, J., series under the editorial supervision of, *Aux confins de l'Univers* [On the borders of the Universe*], Fayard-Fondation Diderot, 1987.

STURROCK, P.A., MOLZER, T.E., MIHALAS, D.M., and ULRICH, R.K., series published by, *Physics of the Sun*, 3 vol., Dordrecht, Reidel, 1986.

---

* These references have not been published in English.